Overheating

Overheating

An Anthropology of Accelerated Change

Thomas Hylland Eriksen

First published 2016 by Pluto Press
345 Archway Road, London N6 5AA

www.plutobooks.com

British Library Cataloguing in Publication Data
A catalogue record for this book is available from the British Library

ISBN 978 0 7453 3639 8 Hardback
ISBN 978 0 7453 3634 3 Paperback
ISBN 978 1 7837 1984 6 PDF eBook
ISBN 978 1 7837 1986 0 Kindle eBook
ISBN 978 1 7837 1985 3 EPUB eBook

Printed and bound by CPI Group (UK) Ltd, Croydon, CR0 4YY

Typeset by Stanford DTP Services, Northampton, England

Simultaneously printed in the European Union and United States of America

Contents

Illustrations

Preface

The contemporary world is … too full? Too intense? Too fast? Too hot? Too unequal? Too neoliberal? Too strongly dominated by humans?

All of the above, and more. Ours is a world of high-speed modernity where the fact that things change no longer needs to be explained by social scientists; what comes across as extraordinary or puzzling are instead the patches of continuity we occasionally discover. Modernity in itself entails change, but for decades change was synonymous with progress, and the standard narrative about the recent past was one of improvement and development. Things seemed to be getting better and history had a direction.

In the last few decades, the confidence of the development enthusiasts has been dampened. Modernity and enlightenment did not eradicate atavistic ideologies, sectarian violence and fanaticism. Wars continued to break out. Inequality and poverty did not go away. Recurrent crises with global repercussions forced economists to concede, reluctantly, at least when caught with their pants down, that theirs was not a precise science after all. Although many countries were democratic in name, a growing number of people felt that highly consequential changes were taking place in their lives and immediate surroundings without their having been consulted beforehand. And, most importantly, the forces of progress turned out to be a double-edged sword. What had been our salvation for 200 years, namely inexpensive and accessible energy, was about to become our damnation through environmental destruction and climate change.

It is chiefly in this sense that it is meaningful to talk of our time as being postmodern. The old recipes for societal improvement, whether socialist, liberal or conservative, have lost their lustre. The political left, historically based on demands for social justice and equality, is now confronted with two further challenges in the shape of multiculturalism and climate change, and creating a consistent synthesis of the three is not an easy task. Generally speaking, in complex systems, the unintended consequences are often more conspicuous than the planned outcomes of a course of action.

This book is based on the assumption that the rapid changes characterising the present age have important, sometimes dramatic, unintended consequences. Each of the five empirical chapters focuses on one key area – energy, mobility, cities, waste, information – and shows how changes may take unexpected directions, which were neither foreseen nor desired at the outset.

Just as the insecticide called DDT – which was meant to save crops and improve agricultural output – killed insects, starved birds and led to 'the silent spring' of Rachel Carson's eponymous book, a foundational text for the modern environmental movement (Carson 1962), so does the car lead to pollution and accidents, the information revolution to the pollution of brains and, perhaps, the spread of Enlightenment ideas leads to counter-reactions in the form of fundamentalism. In the coming chapters, I focus on such contradictions, but I also show that the crises of globalisation are not caused by malevolent intentions or any kind of evil, selfish or short-sighted conspiracy. Rather, what we are confronted with is a series of *clashing scales* which remain poorly understood. Let me give a brief illustration. If you are in a powerful position, you can change thousands of people's lives far away with a stroke of a pen; but if you spent time with them first, that is likely to influence your decision. The tangibly lived life at the small scale, in other words, clashes with large-scale decisions, and you come to realise that what is good for Sweden is not necessarily good for the residents of the village of Dalby north of Ystad.

Scaling up can be an efficient way of diverting attention from the actuality of a conflict by turning it into an abstract issue. If your colleagues complain that you never make coffee for your co-workers, you may respond, scaling up a notch, that the neoliberal labour regime is so stressful and exhausting that the ordinary office worker simply has no time for such luxuries. At the opposite end of the spectrum, we may think about the workers who manned the gas chambers and effectively murdered incomprehensible numbers of Jews and Gypsies; there is no indication that they loved their family and household pets less than anyone else did. I will show how different the world, or an activity, or an idea, looks when you move it up and down the scales. As one of my informants in an industrial Australian city said: 'The environmental activists in Sydney are really good at saving the world, but they don't have a clue as to what to do with real people with factory jobs.'

Being an anthropologist and, accordingly, trained to seeing the world from below, I have often had mixed feelings about the general literature about globalisation. Many widely read authors writing about the interconnected world seem to be hovering above the planet in a helicopter with a pair of binoculars. They may get the general picture right, but fail to see the nooks and crannies where people live. Reading these books, I am always reminded of Benoît Mandelbrot's article 'How long is the coast of Britain?' (Mandelbrot 1967), which is fundamentally about scaling. He shows that the length of the jagged British coast depends on the scale of the map. Measuring with a yardstick would produce a different result from measurements taken with a one-foot ruler. And in order to get to the truth about people's lives, the bird's eye perspective is useful, but inadequate. You have to get 'up close and personal' (Shore and Trnka 2013).

If you read general overviews about globalisation and identity with the mindset of an anthropologist, there is a chance that you end up with the somewhat unsatisfactory feeling that you had been offered a three-course dinner, and were duly served a sumptuous starter and a delicious dessert, but no main course. With anthropologists, the problem is generally the opposite: They describe local life-worlds in meticulous detail, crawling, as it were, on all fours with a magnifying glass, but rarely attempt a global analysis. By moving up and down the scales, I shall try to do both, and to relate them to each other.

This is a book about matters of great concern to humanity: climate and the environment, urbanisation and improvisational survivalism, tourism and migration, waste and inequality, excess and deprivation. In an attempt to account for the recent failure of the standard modern narrative about development and progress, I look for paradoxes, contradictions, clashes of scale and runaway processes. Both the questions and the concepts are big. Yet this book is modest in size. One reason is that it only marks a beginning. Growing out of the research project 'Overheating: The three crises of globalisation', funded by the European Research Council, this is an overture and an overview. It is meant as a stand-alone publication but, at the same time, it introduces topics, concepts and approaches which will be developed in greater empirical detail and theoretical depth in later publications. *Overheating* consists of a series of interrelated ethnographic projects which aim to produce comparable and compatible data on the local perception, impact and management of the global crises. In this way, both the myopic bias of anthropology and the top-down approach of other social sciences are transcended. The individual 'Overheating' projects are scattered across the planet, but they speak to larger issues of global importance as well as maintaining an ongoing conversation with each other, through commitment to a number of shared presuppositions, research questions and concepts.

It might be expected that a book about acceleration and globalisation would delve into such phenomena as financialisation and ideologies of culture, from creolisation to identity politics. I might well have done so, but these overheating phenomena will be dealt with properly in later publications. Instead, I have concentrated on some of the main material dimensions of overheating: energy use, with a particular emphasis on fossil fuels; human mobility and immobility as a simultaneous cause and effect of overheating; the growth of cities and peri-urban areas, a result partly of privatisation of land rights and population growth in the Global South; the mountains and rivers of waste that humans can no longer leave behind, but are instead embedded in, at the height of the Anthropocene; and the incredibly fast development and spread of wireless information technology. Granted, there are other material dimensions of overheating that could have been dealt with here. Trade and transport are perhaps the most obvious ones, with a possible focus on the container ship as a precondition for global neoliberalism and the rise of China as a world leader in commodity exports (Levinson 2006; Bear 2014). In my defence, I should point out that an investigation of this trend (like several others, such as the global mining boom or infrastructure development, for example roads) would mainly add quantitatively to the analysis without enriching it conceptually. The standardisation of shipping containers and ports, ships and railways adapted to them is an essential condition for overheated neoliberalism. Container-ship transportation has grown exponentially since 1960, leading to a massive clash of scales (between large-scale trade and small-scale production in importing or competing locations) and implying a loss of flexibility through its reliance on plantation-style monoculture (cf. Tsing 2012) in the production and distribution regimes. I will nevertheless return to the overheating effects of the container ship, roads and the information revolution in the final chapter, which offers an integrated perspective on runaway globalisation – overheating – seen through the lens of scale.

I should finally warn the reader that this is not a typical academic book. It is written in non-technical language, and it is eclectic in its theorising, catholic in its use of sources, which range from the newspaper article to the historical treatise. There are many ways of knowing, and they are often complementary rather than contradictory. Although the ethnographic gaze, which privileges the local life-world is important, it has to be contextualised both historically and with large-scale perspectives.

The core concepts of the book are scale and scaling, double bind, runaway processes and treadmill competition, flexibility and reproduction. The main analytical focus is on the clashes of scales in a situation of runaway growth where an increasing number of humans are beginning to understand the impossibility of continued growth and the severe side-effects of that growth. This is a world where the number of photos taken is estimated to have trebled between 2010 and 2015; a world where the largest container port in the world (Shanghai) grew from next to nothing in 1990 to 11.4 million TEUs annually in 2003 and on to 35 million TEUs in 2014 (a TEU is a standard 20-foot container); a world where a fast growing number of people live in cities, go on holiday, drive cars and throw away their rubbish to an extent which is not only unprecedented, but which was difficult to imagine as recently as the late twentieth century. The graphs, numbers and statistics scattered around the text are therefore meant as indications of a tendency, not as facts to be remembered in detail. It is in the nature of the changes we currently witness that they will be outdated very soon. The argument, hopefully, will not.

* * *

The work involved in the writing of this book has been funded by an Advanced Grant from the European Research Council (ERC), entitled 'Overheating: The three crises of globalisation, or An anthropological history of the early 21st century'. I am grateful to the ERC for having given me the opportunity to think in bigger and broader terms than is usually possible in the contemporary academic world. I am also grateful to my 'Overheating' colleagues – Cathrine Moe Thorleifsson, Elisabeth Schober, Robert Pijpers, Lena Gross, Henrik Sinding-Larsen, Wim Van Daele, Astrid Stensrud and Chris Hann, as well as our excellent MA students – for their many queries, criticisms, suggestions and comments on the first draft and for constructive criticism of my ideas in general.

Oslo
March 2016

1. *Le monde est trop plein*

On 28 November 2008, the celebrated French intellectual Claude Lévi-Strauss, the founder of structuralism, celebrated his hundredth birthday. He had been one of the most important anthropological theorists of the twentieth century, and, although he had ceased publishing years ago, his mind had not yet given in. But his time was nearly over, and he knew it. The book many consider his most important had been published almost 60 years earlier. When *Les Structures élémentaires de la parenté* (*The Elementary Structures of Kinship*, Lévi-Strauss 1969 [1949]) appeared, it transformed anthropological thinking about kinship by shifting the focus of the field and reconceptualising this most universal of all social modes of being. To Lévi-Strauss, the most significant fact about kinship was not descent from a common ancestor, but rather the alliances between groups created by marriage.

On his birthday, Lévi-Strauss received a visit from President Nicolas Sarkozy, since France remains a country where politicians can still increase their symbolic capital by socialising with intellectuals. During the brief visit, the ageing anthropologist remarked that he scarcely considered himself among the living any more. By saying so, he did not merely refer to his advanced age and weakened capacities, but also to the fact that the world to which he had devoted his life's work was by now all but gone. The small, stateless peoples who had featured in most of his world had by now been incorporated, with or against their will, into states, markets and monetary systems of production and exchange.

During his brief conversation with the president, Lévi-Strauss also remarked that the world was too full: *Le monde est trop plein*. By this he clearly referred to the fact that the world was filled by people, their projects and the material products of their activities. The world was *overheated*. There were by now 7 billion of us compared to 2 billion at the time of the great French anthropologist's birth, and quite a few of them seemed to be busy shopping, posting updates on Facebook, migrating, working in mines and factories, learning the ropes of political mobilisation or acquiring the rudiments of English.

Lévi-Strauss had bemoaned the disenchantment of the world since the beginning of his career. Already in his travel memoir *Tristes Tropiques*, published in 1955, he complained that:

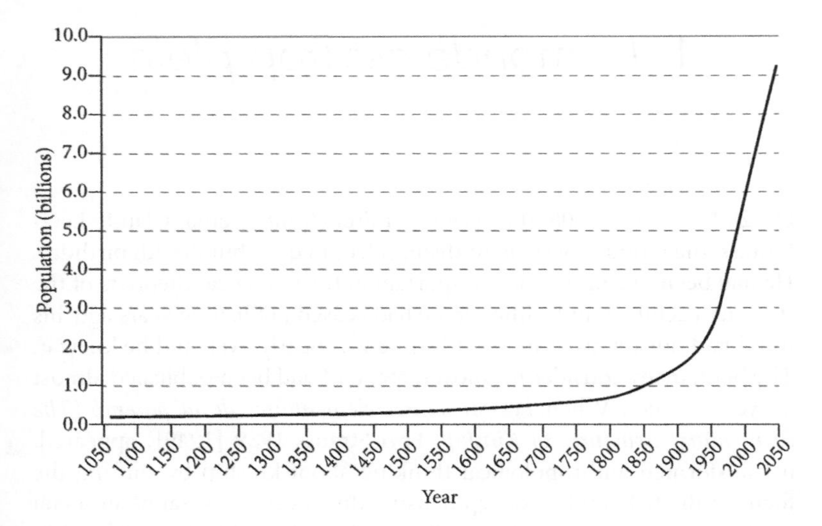

Figure 1.1 World population growth since 1050

Source: Puiu (2015).

> [n]ow that the Polynesian islands have been smothered in concrete and turned into aircraft carriers solidly anchored in the southern seas, when the whole of Asia is beginning to look like a dingy suburb, when shanty towns are spreading across Africa, when civil and military aircraft blight the primaeval innocence of the American or Melanesian forests even before destroying their virginity, what else can the so-called escapism of travelling do than confront us with the more unfortunate aspects of our history? (Lévi-Strauss 1961 [1955]: 43)

– adding, famously, with reference to the culturally hybrid but undeniably modern people of the cities in the New World, that they had taken the journey directly from barbarism to decadence without passing through civilisation. The yearning for a lost world is evident, but anthropologists have been nostalgic far longer than this. Ironically, the very book which would change the course of modern European social anthropology more than any other, conveyed pretty much the same message of loss and nostalgia. Bronislaw Malinowski's *Argonauts of the Western Pacific*, published just after the First World War, begins with the following prophetic words:

> Ethnology is in the sadly ludicrous, not to say tragic, position, that at the very moment when it begins to put its workshop in order, to

> forge its proper tools, to start ready for work on its appointed task, the material of its study melts away with hopeless rapidity. Just now, when the methods and aims of scientific field ethnology have taken shape, when men [sic] fully trained for the work have begun to travel into savage countries and study their inhabitants – these die away under our very eyes. (1984 [1922]: xv)

Disenchantment and disillusion resulting from the presumed loss of radical cultural difference has, in a word, been a theme in anthropology for a hundred years. It is not the only one, and it has often been criticised, but the Romantic quest for authenticity still hovers over anthropology as a spectre refusing to go away. Clifford Geertz and Marshall Sahlins, the last major standard-bearers of classic cultural relativism, each wrote an essay in the late twentieth century where they essentially concluded that the party was over. In 'Goodbye to tristes tropes', Sahlins quotes a man from the New Guinea Highlands who explains to the anthropologist what *kastom* (custom) is: 'If we did not have *kastom*, we would be just like the white man' (Sahlins 1994: 378). Geertz, for his part, describes a global situation where 'cultural difference will doubtless remain – the French will never eat salted butter. But the good old days of widow burning and cannibalism are gone forever' (Geertz 1986: 105).

It is a witty statement, but it is nonetheless possible to draw the exact opposite conclusion. Regardless of the moral position you take, faced with the spread – incomplete and patchy, but consequential and important – of modernity, it is necessary to acknowledge, once and for all, that mixing, accelerated change, connectedness and the uneven spread of modernity is the air that we breathe in the present world. Moreover, we may argue that precisely because the world is *trop plein*, full of interconnected people and their projects, it is an exciting place to study right now. People are aware of each other in ways difficult to imagine only a century ago; they develop some kind of global consciousness and often some kind of global conscience virtually everywhere. Yet their global outlooks remain firmly anchored in their worlds of experience, which in turn entails that there are many distinctly local global worlds.

People now build relationships which can just as well be transnational as local, and we are connected through the increasingly integrated global economy, the planetary threat of climate change, the hopes and fears of virulent identity politics, consumerism, tourism and media consumption. One thing that it is not, incidentally, is a homogenised world society where everything is becoming the same. Yet, in spite of the differences and inequalities defining the early twenty-first-century world, we are slowly learning to take part in the same conversation about humanity and where it is going.

In spite of its superior research methods and sophisticated tools of analysis, anthropology struggles to come properly to terms with the world today. It needs help from historians, sociologists and others. The lack of historical depth and societal breadth in anthropology has already been mentioned, and a third problem concerns normativity and relativism. For generations, anthropologists were, as a rule, content to describe, compare and analyse without passing moral judgement. The people they studied were far away and represented separate moral communities. Indeed, the method of cultural relativism requires a suspension of judgement to be effective. However, as the world began to shrink as a result of accelerated change in the postwar decades, it increasingly became epistemologically and morally difficult to place 'the others' on a different moral scale than oneself. The *de facto* cultural differences also shrank as peoples across the world increasingly began to partake in a bumpy, unequal but seamless global conversation. By the turn of the millennium, tribal peoples were rapidly becoming a relic, although a dwindling number of tribal groups continue to resist some of the central dimensions of modernity, notably capitalism and the state. Indigenous groups have become accustomed to money, traditional peasants' children have started to go to school, Indian villagers have learned about their human rights, and Chinese villagers have been transformed into urban industrial workers. In such a world, pretending that what anthropologists did was simply to study remote cultures, would not just have been misleading, but downright disingenuous.

The instant popularity of the term 'globalisation' coincided roughly with the fall of the Berlin wall, the beginning of the end of apartheid, the coming of the internet and the first truly mobile telephones. This world of 1991, which influences and is being influenced by different people (and peoples) differently and asymmetrically, rapidly began to create a semblance of a global moral community where there had formerly been none, at least from the viewpoint of anthropology. Ethnographers travelling far and wide now encountered indigenous Amazonian people keen to find out how they could promote their indigenous rights in international arenas, Australian aborigines poring over old ethnographic accounts in order to relearn their half-forgotten traditions, Indian women struggling to escape from caste and patriarchy, urban Africans speaking cynically about corrupt politicians and Pacific islanders trying to establish intellectual copyright over their cultural production in order to prevent piracy.

In such a world, the lofty gaze of the anthropological aristocrat searching for interesting dimensions of comparison comes across not only as dated but even as somewhat tasteless. Professed neutrality becomes in itself a political statement.

What had happened – apart from the fact that native Melanesians now had money, native Africans mobile phones and native Amazonians rights claims? The significant change was that the world had, almost in its entirety, been transformed into a single – if bumpy, diverse and patchy – moral space, while the anthropologists had been busy looking the other way.

In this increasingly interconnected world, cultural relativism can no longer be an excuse for not engaging with the victims of patriarchal violence in India, human rights lawyers in African prisons, minorities demanding not just cultural survival but fair representation in their governments. Were one to refer to 'African values' in an assessment of a particular practice, the only possible follow-up question would be '*whose* African values'? In this world, there is friction between systems of value and morality. There can be no retreat into the rarefied world of radical cultural difference when, all of a sudden, some of the 'radically culturally different' ask how they can get a job, so that they can begin to buy things. The suture between the old and the new can be studied by anthropologists, but it must be negotiated by those caught on the frontier, and in this world, the anthropologist, the 'peddler of the exotic' in Geertz's (1986) words, cannot withdraw or claim professional immunity, since the world of the remote native is now his own.

Contradictions

Many useful and informative books have been written on globalisation since around 1990. Some of them highlight contradictions and tensions within the global system that are reminiscent of the dialectics of globalisation as described here – George Ritzer (2004) speaks of 'the grobalization of nothing' (his term 'grobalization' combines growth and globalisation) and 'the glocalization of something', Manuel Castells (1996) about 'system world' and 'life world' (in a manner akin to Niklas Luhmann), Keith Hart (2015) contrasts a human economy with a neoliberal economy, and Benjamin Barber (1995) makes a similar contrast with his concepts of 'Jihad' and 'McWorld'. In all cases, the local strikes back at the homogenising and standardising tendencies of the global.

The extant literature on globalisation is huge, but it has its limitations. Notably, most academic studies and journalistic accounts of global phenomena tend to iron out the unique and particular of each locality, either by treating the whole world as if it is about to become one huge workplace or shopping mall, and/or by treating local particularities in a cavalier and superficial way. The anthropological studies that exist of globalisation, on the other hand, tend to limit themselves to one or a few aspects of globalisation, and to focus too exclusively on exactly that local reality which the more wide-ranging studies neglect. These limitations must be transcended dialectically, by building the confrontation between the universal and the particular into the research design as a premise: For a perspective on the contemporary world to be convincing and comprehensive, it needs the view from the helicopter circling the world just as much as it needs the details that can only be discovered with a magnifying glass. The macro and the micro, the universal and the particular must be seen as two sides of the same coin. One does not make sense without the other; it is yin without yang, Rolls without Royce.

In order to explore the local perceptions and responses to globalisation, no method of inquiry is superior to ethnographic field research. Unique among the social science methods, ethnography provides the minute detail and interpretive richness necessary for a full appreciation of local life. This entails a full understanding of local interpretations of global crisis and their consequences at the level of action. Moreover, there is no such thing as *the* local view. Within any community views vary since people are differently positioned. Some gain and some lose in a situation of change; some see loss while others see opportunity. But none can anticipate the long-term implications of change.

While ethnography is the richest and most naturalistic of all the social science methods, it is not sufficient when the task at hand amounts to a study of global interconnectedness and, ultimately, the global system. The methods of ethnography must therefore be supplemented. Ethnography can be said to be enormously deep and broad in its command of human life-worlds, but it can equally well be said that it lacks both depth and breadth, that is historical depth and societal breadth. A proper grasp of the global crises, in other words, requires both a proper command of an ethnographic field and sufficient contextual knowledge – statistical, historical, macrosociological – to allow ethnography to enter into the broad conversation about humanity at the outset of the twenty-first century. Since human lives are lived in the concrete here and now, not as abstract generalisations, no account of globalisation is complete unless it is anchored in a local life-world – but understanding local life is also in itself inadequate, since the local reality in itself says little about the system of which it is a part.

It is only in the last couple of decades that the term 'globalisation' has entered into common usage, and it may be argued that capitalism, globally hegemonic since the nineteenth century, is now becoming universal in the sense that scarcely any human group now lives independently of a monetised economy. Traditional forms of land tenure are being replaced by private ownership, subsistence agriculture is being phased out in favour of wage work, TV replaces orally transmitted tales and, since 2007, UN estimates suggest that more than half the world's population lives in urban areas (expected to rise to 70 per cent by 2050). The state, likewise, enters into people's lives almost everywhere, though to different degrees and in different ways.

It is an interconnected world, but not a smoothly and seamlessly integrated one. Rights, duties, opportunities and constraints continue to be unevenly distributed, and the capitalist world system itself is fundamentally volatile and contradiction-ridden, as its recurrent crises, which are rarely predicted by experts, indicate. One fundamental contradiction consists in the chronic tension between the universalising forces of global modernity and the desire for autonomy in the local community or society. The drive to standardisation, simplification and universalisation is always countered by a defence of local values, practices and relations. In other words, globalisation does not lead to global homogeneity, but highlights a tension, typical of modernity, between the system world and the life-world, between the standardised and the unique, the universal and the particular.

At a higher level of abstraction, the tension between economic development and human sustainability is also a chronic one, and it constitutes the most fundamental double bind of twenty-first-century capitalism. Almost everywhere, there are trade-offs between economic growth and ecology. There is a broad global consensus among policy makers and researchers that the global climate is changing irreversibly due to human activity (mostly the use of fossil fuels). However, other environmental problems are also extremely serious, ranging from air pollution in cities in the Global South to the depletion of phosphorus (a key ingredient in chemical fertiliser), overfishing and erosion. Yet the same policy makers who express concern about environmental problems also advocate continued economic growth, which so far has presupposed the growing utilisation of fossil fuels and other non-renewable resources, thereby contradicting another fundamental value and contributing to undermining the conditions for its own continued existence.

This globally interconnected world may be described through its tendency to generate chronic crises, being complex in such a way as to be ungovernable, volatile and replete with unintended consequences – there are double binds, there is an uneven pace of change, and an unstable relationship between universalising and localising processes. Major transformations engendered by globalisation are those relating to the environment, of the economy, and of identity. They are interconnected and relatively autonomous, although the fundamental contradiction in the global system, arguably, is the conflict between growth and sustainability; these three crises share key features, and they are perceived, understood and responded to locally across the world.

The perspective I am developing concerns these transformations. It represents a critical perspective on the contemporary world since it insists on the primacy of the local and studies global processes as inherently contradictory. I also aim to make a modest contribution to an interdisciplinary history of the early twenty-first century with a basis in ethnography. By this, I mean that it is misleading to start a story about the contemporary world by looking at the big picture – the proportion of the world's population who are below the UN poverty limit; the number of species driven to extinction in the last half-century; the number of internet users in India and Venezuela – unless these large, abstract figures are related to people's actual lives. It is obvious that 7 per cent economic growth in, say, Ethiopia does not automatically mean that all Ethiopians are 7 per cent better off (whatever *that* means), yet those who celebrate abstract statistical figures depicting economic growth often fail to look behind the numbers. They remain at an abstract level of scale, which is not where life takes place. Similarly, the signing of an international agreement on climate change, which took place in Paris in December 2015, does not automatically entail practices which mitigate climate change. So while trying to weave the big tapestry and connect the dots, the credibility of this story about globalisation depends on its ability to show how global processes interact with local lives, in ways which are both similar and different across the planet.

This is a story of contemporary neoliberal global capitalism, the global information society, the post-Cold War world: The rise of information technologies enabling fast, cheap and ubiquitous global communication in real time, the demise of 'the Second World' of state socialism, the hegemony of neoliberal economics, the rise of China as an economic world power, the heightened political tensions, often violent, around religion (especially Islam, but also other religions), the growing concern for the planet's ecological future in the political mainstream, and the development of a sprawling, but vocal 'alterglobalisation movement' growing out of discontent with the neoliberal world order – all these recent and current developments indicate that this is indeed a new world, markedly different from that of the twentieth century which, according to a widespread way of reckoning (for example Hobsbawm 1994), began with the First World War and the Russian Revolution, and ended with the final dissolution of the Soviet Union in 1991.

The globalisation discourse tends to privilege flows over structures, rhizomes over roots, reflexivity over doxa, individual over group, flexibility over fixity, rights over duties, and freedom over security in its bid to highlight globalisation as something qualitatively new. While this kind of exercise is often necessary, it tends to become one-sided. Anthropologists may talk disparagingly about the jargon of 'globalbabble' or 'globalitarism' (Trouillot 2001), and tend to react against simplistic generalisations by reinserting (and reasserting) the uniqueness of the local, or the glocal if you prefer.

There is doubtless something qualitatively new about the compass, speed and reach of current transnational networks. Now, some globalisation theorists argue that the shrinking of the world will almost inevitably lead to a new value orientation, some indeed heralding the coming of a new, postmodern kind of person (for example Sennett 1998). These writers, who predict the emergence of a new set of uprooted, deterritorialised values and fragmented identities, are often accused of generalising from their own European middle-class habitus, the 'class consciousness of frequent travellers' (Calhoun 2002). The sociologist John Urry, who may be seen as a target for this criticism, argues in *Global Complexity* (2003) that globalisation has the potential of stimulating widespread cosmopolitanism – however, he does not say among whom. At the same time, Urry readily admits that the principles of closeness and distance still hold in many contexts, for example in viewing patterns on television, where a global trend consists in viewers' preferences for locally produced programmes.

The newness of the contemporary world was described by Castells in his trilogy *The Information Age* (1996, 1997, 1998), where – after offering a smorgasbord of new phenomena, from real-time global financial markets to the spread of human rights ideas – he remarks, in a footnote tucked away towards the end of the third and final volume, that what is new and what is not does not really matter; his point is that 'this is our world, and therefore we should study it' (Castells 1998: 336). However much I appreciate Castells' analysis, I disagree. It does matter what is new and what isn't if we are going to make sense of the contemporary world. Different parts of societies, cultures and life-worlds change at different speeds and reproduce themselves at different rhythms, and it is necessary to understand the disjunctures between speed and slowness, change and continuity in order to grasp the conflicts arising from accelerated globalisation.

Accelerated growth

The first fact about the contemporary world is accelerated growth. There are more of us, we engage in more activities, many of them machine-assisted, and depend on each other in more ways than ever before. No matter how you go about measuring it, it is impossible not to conclude that connectedness and growth have increased phenomenally. There are more of us than at any earlier time, and each of us has, on an average, more links with the outside world than our parents or grandparents. We have long been accustomed to the steep curves depicting world population growth, but the fastest growth does not take place in the realm of population. It stands to reason that the number of people with access to the internet has grown at lightning speed since 1990, since hardly anyone was online at the time. But the growth in internet use continues to accelerate. Only in 2006, it was estimated that less than 2 per cent had access to the internet in sub-Saharan Africa (bar South Africa, which has a different history). By 2016, the percentage is estimated to be between 25 and 30 per cent, largely owing to affordable smartphones rather than a mushrooming of internet cafes or the spread of laptops among Africans. Or we could look at migration. When, around 1990, I began to write about cultural diversity in my home country, Norway, there were about 200,000 immigrants (including first-generation descendants) in the country. By 2016, the figure exceeds 800,000. Or we could look at urbanisation in the Global South. Cities like Nouakchott in Mauritania and Mogadishu in Somalia have grown, since the early 1980s, from a couple of hundred thousand to a couple of million inhabitants each. The growth has been 1000 per cent in one generation.

Or we could take tourism. As early as the 1970s, cultured North Europeans spoke condescendingly of those parts of the Spanish coast that they deemed to have been 'spoiled' by mass tourism. In 1979, shortly after the end of Fascism in the country, Spain received about 15 million tourists a year. In 2015, the number was about 60 million. We are, in other words, talking about a fourfold growth in less than 40 years.

The growth in international trade has been no less spectacular than that in tourism or urbanisation. The container ship with its associated cranes, railways, standardised metal containers and reconstructed ports, perhaps the symbol *par excellence* of an integrated, standardised, connected world (Levinson 2006), slowly but surely gained importance from its invention in the 1950s until it had become the industry standard a few decades later. The ports of Shanghai and Singapore more than doubled their turnover of goods between 2003 and 2014. While world GDP is estimated to have grown by 250 per cent since 1980, world trade grew 600 per cent in the same period (IMF), a development made possible not least through the reduced transport costs enabled by the shipping container.

* * *

Websites, international organisations, conferences and workshops, mobile phones and TV sets, private cars and text messages: the growth curves point steeply upwards in all these – and many other – areas. In 2005, Facebook did not exist; a decade later, the platform had more than a billion users.

Not all change accelerates, and not everything that changes has similarly momentous consequences. Although the growth in tourism has been staggering, it has been slower than the growth in text messages. But although phenomena like text messaging and Facebook, tourism and cable TV have transformed contemporary lives in ways we only partly understand, there are two changes of a material nature which are especially relevant for an understanding of the contemporary world, and which have undisputable consequences for the future: Population growth and the growth in energy use.

The growing human population of 7 billion (compared to 1 billion in 1800 and 2 billion as late as 1920) travels, produces, consumes, innovates, communicates, fights and reproduces in a multitude of ways, and we are increasingly aware of each other as we do so. The steady acceleration of communication and transportation in the last two centuries has facilitated contact and made isolation difficult, and is weaving the growing global population ever closer together, influencing but not erasing cultural differences, local identities and power disparities. Since we are now seven times as many as we used to be at the end of the Napoleonic wars, it comes as no surprise that we use more energy today; but the fact is that energy use in the world has grown much faster than the world population. In 1820, each human used on an average 20 Gigajoules a year. Two centuries later, we have reached 80, largely thanks to the technology that enabled large-scale use of fossil fuels. Consumption is far from evenly distributed, and so those of us who live in rich countries have access to so much machine power that it can be compared to having 50 slaves each.

The quadrupling in energy use is in reality a growth by a factor of 28, since there are seven times as many of us today as in 1814. The side-effects are well known. The visible and directly experienced ones are pollution and environmental degradation. Effects which are both more difficult to observe and more consequential are long-term climate changes and the depletion of (non-renewable) energy sources.

If population had not begun to grow exponentially in the nineteenth century, humanity might have evaded the most serious side-effects of the fossil fuel revolution. The original thinker and maverick scientist James Lovelock writes, in his *Revenge of Gaia* (Lovelock 2006), that had there just been a billion of us, we could probably have done as we liked. The planet would recover. Similarly, it is possible to imagine, although the scenario is unrealistic, that world population could have increased sevenfold without the fossil fuel revolution. In that case, the climate crisis would have been avoided, but the great majority of the world would have lived in a state of constant, abject poverty. Instead, we now live in a world where modernity has shifted to a higher gear, where there is full speed ahead in most areas. It is a volatile and ultimately self-destructive situation. In addition, continued growth is impossible, as shown by Jeremy Grantham (Grantham 2011; see also Hornborg 2011; Monbiot 2014; Rowan 2014). This is a central conundrum of contemporary modernity, making conventional Enlightenment–industrial ideas of progress and development far more difficult to defend now than just a generation ago.

The question is: what could be a feasible alternative in a world society which seems to have locked itself to a path which is bound to end with collapse? There is no simple, or single, answer to this, the most important question of our time. Indeed, there is not even general agreement about how to phrase the question. Healthy doses of intellectual and political imagination will be necessary to move ahead, and one size does not fit all. As pointed out by the economist Elinor Ostrom, famous for showing how communities are capable of managing resources sustainably, there is no reason to assume that what works in Costa Rica will work in Nepal (Ostrom 1990). Each place is interwoven with every other place, but they also remain distinctive and unique.

The acceleration of history

While some optimistic liberals argued, in a pseudo-Hegelian way, that the destruction of the Berlin wall, would mean the end of history, this did not occur. To the contrary, history accelerated (Hann 1994) but without a clear direction.

As late as the 1970s, there existed a widely shared hegemonic narrative about the way in which the modern world had grown. It was a story about enlightenment and invention, conquest and decolonisation, progress and welfare, war and peace. The story existed in loyal and critical versions; it was produced in liberal, conservative, socialist or communist flavours. And then, the story of progress lost its lustre, not with a bang, but with a whimper. As of today, there is no story about where we are coming from and where we are going with general appeal in most, or even any, part of the world. Perhaps changes are taking place too fast – it has been said that people belonging to the global middle class today experience 17 times as much as their great-grandparents, but without an improved apparatus for digesting and understanding their experiences. Or perhaps the side-effects of progress have simply become too noticeable. We have at our disposal lucid, didactically clear stories about the transition from hunting and gathering to agriculture and on to the Industrial Revolution. Naturally, they are contested by feminists, postcolonials, Marxists, post-structuralists and many others, and they are continuously being rewritten, but the template is there. No similar story exists about the global information society and where it is headed. Or, rather, if they are at all considered, future prospects appear to be bleak, based on what the sociologist Frank Furedi (2002) calls 'a culture of fear'.

If periodisation is necessary; if we need a date for the transition from modernity to postmodernity, I suggest 1991, a year to which I have already alluded. First, 1991 was the year in which the Cold War ended in its original form. The two-bloc system that had defined the postwar period was suddenly gone. The ideological conflict between capitalism and socialism finally seemed to have been replaced by the triumphant sound of one hand clapping. At the same time, the Indian economy was deregulated massively by Rajiv Gandhi's government. By 1991, it was also clear that apartheid was about to be relegated to the dustbin of history. Mandela had been released from prison the year before, and negotiations between the Nationalist Party and the ANC (African National Congress) had begun in earnest. The future of the entire world (notwithstanding a few stubborn outliers like Cuba and North Korea) seemed to consist in a version of global neoliberalism – that is, a virulent and aggressive form of deregulating capitalism where the main role of the state consisted in ensuring the functioning of so-called free markets. However, it soon became clear that neoliberalism did not deliver the goods. Social inequalities continued to exist, and in some countries, like the US, they grew enormously. Countries in the Global South did not develop along the predicted lines, that is, roughly in the same way as the countries of the Global North. Influential commentators such as the economist Joseph Stiglitz (2002), the investor George Soros (2002) and the philosopher John Gray (1998), former supporters of the neoliberal paradigm, changed their minds and wrote scathing critiques of the deregulated global economy. At the same time, politicised religion and other forms of identity politics flourished from India to Israel, from Belfast to Baluchistan, contrary to predictions that education and modernity would weaken such forms of political identity, which were often divisive and reactionary in character. The war in Yugoslavia and the Rwandan genocide, both unfolding in the mid 1990s, were reminders that an identity based on notions of kinship and descent did not belong to the past, but remained important for millions, and could erupt in horrible ways at any time. The year 1991 was also the height of the Salman Rushdie affair. Rushdie's novel *The Satanic Verses* had been published in 1988, denounced as blasphemous by powerful Muslim clerics, and the author had been sentenced to death *in absentia* by Iranian clergy. The affair was a tangible reminder of a new kind of interconnectedness, where local or domestic acts can have instant global ramifications.

Around the same time, mobile telephones and the internet began to spread epidemically in the global middle classes, eventually trickling down to the poor as well. A certain kind of flexibility grew: you could soon work anywhere and at any time, but these technologies contributed to fragmentation as well; what flexibility was gained with respect to space seemed to be lost regarding time. Life began to stand still at a frightful speed. Your gaze was now fixed at a point roughly one minute ahead. This spelled bad news for the slow, cumulative time of growth and development (Eriksen 2001; Morozov 2012).

A similar kind of flexibility began to affect labour and business, and it was not the kind of flexibility that offers alternative paths for action, but one which created insecurity and uncertainty. Companies that used to distinguish between short-term and long-term planning ceased to do so, since everything now seemed to be short term; nobody knows what the world may look like in five or ten years' time. To workers, the most perceptible change is basic insecurity. One of the most widely used new concepts in the post-millennial social sciences (along with the Anthropocene and neoliberalism) is *the precariat* (Standing 2011), and there are good reasons for its sudden popularity. The precariat consists of the millions of employees whose jobs are short-term and temporary, and who, accordingly, have no clue as to whether they will have work next year or even next month. This 'new dangerous class', as the economist Guy Standing has it, is as easily found in the British construction sector and in Danish universities as in a Mexican sweatshop or a shipyard in the Philippines. To millions of people, the freedom of neoliberal deregulation merely means insecurity and reduced autonomy.

It is a new world, but not in every relevant respect. The task at hand now, before going into some facts and stories about specific overheating effects, consists in finding the appropriate vocabulary for talking about the new and the old, that which changes and that which does not, the horizontal linkages and the vertical ones. The next chapter introduces a conceptual inventory, or framework, which I have found helpful in trying to make sense of this overheated world.

2. A Conceptual Inventory

The accelerated and intensified contact which is a defining characteristic of globalisation leads to tensions, contradictions, conflict and changed opportunities in ways that influence everyday life as well as large-scale processes in all parts of the world. Change takes place unevenly, and often as a result of a peculiar combination of local and transnational processes, with unpredictable small-scale results from large-scale events. Such forms of change lead to overheating effects in local settings worldwide: unevenly paced change where exogenous and endogenous factors combine to lead to instability, uncertainty and unintended consequences in a broad range of institutions and practices, and contribute to a widely shared feeling of powerlessness and alienation. People perceive, understand and act upon the changes in widely differing ways, depending on their position in their local community and on the characteristics of the locality, as well as its position within regional, national and transnational systems. In order to understand globalisation, it is necessary to explore how its crises are being dealt with in local contexts – how people resist imposed changes, negotiate their relationship to global and transnational forces – and what strategies for survival, autonomy and resistance are being developed. These explorations must take the *genius loci* of the locality seriously, situate the locality historically and connect it to an analysis of global processes. Finally, in order to demonstrate the ubiquity of overheating effects, systematic comparison between otherwise very different localities is necessary.

Locally, the transformations are perhaps best understood as challenges to reproduction: people across the world have to renegotiate the ways in which they sustain themselves economically; their right to define who they are is under pressure, sometimes resulting in crises of identity; and the physical environment changes in ways which sometimes indicate that contemporary world civilisation is ultimately unsustainable. In studying and theorising overheated globalisation, I have found the analytical terms 'double bind', 'flexibility', 'runaway processes' (and 'treadmill syndromes') and 'reproduction', the descriptive terms 'Anthropocene' and 'neoliberalism', and the sorting device of scale, particularly useful. For the sake of subsequent clarity, they will now be introduced and explained.

* * *

Anthropocene. Never before has humanity placed its stamp on the planet in ways even remotely comparable to the situation today. Global human domination is such that the natural scientists Paul Crutzen and Eugene Stoermer proposed, in 2000, naming the current geological era the Anthropocene, based on the realisation that human footprints were now everywhere (Steffen et al. 2007). Even in patches of rainforest or desert where no human has set foot, traces of human activity are present through the local effects of climate change – drought, flooding, the spread of humanly introduced species and so on. If the new nomenclature were to be officially adopted by the International Commission on Stratigraphy, the Holocene (which began just after the last Ice Age, about 12,000 years ago) becomes a very brief interlude in the history of the planet, but the Anthropocene is likely to be much shorter if the growth continues. Be this as it may, we live in an era which, since the onset of the Industrial Revolution in Europe, is marked by human activity and expansion in unprecedented ways.

In the present world, nature has somehow collapsed into culture: Rather than being culture's threatening other, nature is now often portrayed as being fragile and weak, itself threatened by and requiring the protection of caring and compassionate humans aware of their special responsibility for the entire planet. Yet, at the same time, nature seems to strike back at us through violent and unpredictable weather events linked to climate change, droughts and flooding, and, in the longer term, by threatening to make the planet inhospitable to humans. Of course, this is just a manner of talking. 'Nature' is not an acting subject and has no intentions. Yet it is worth noting that human dominance on the planet has many unintended consequences, some very long-term and large-scale.

The now familiar notion of the Anthropocene is a large-scale, indeed global conceptualisation of a set of overheating effects. Since 2010, several new academic and semi-academic journals dedicated to studies of the Anthropocene have been launched. A graph representing the growth of its use since 2010 might well take the same shape as Figure 2.1, which depicts, and suggests connections between, species extinctions and human population growth.

In a less densely populated and intensely exploited world, fish catches were limited by the number of boats in the water, reflecting a scarcity of man-made capital. In today's full world, boats are abundant, but the fish catch is limited by the number of fish remaining in the sea. It is the warehouses of nature that are becoming empty, not human skill or potential that is wanting in efficacy. We humans are overachievers in the planetary context; we succeeded too well in proliferating and

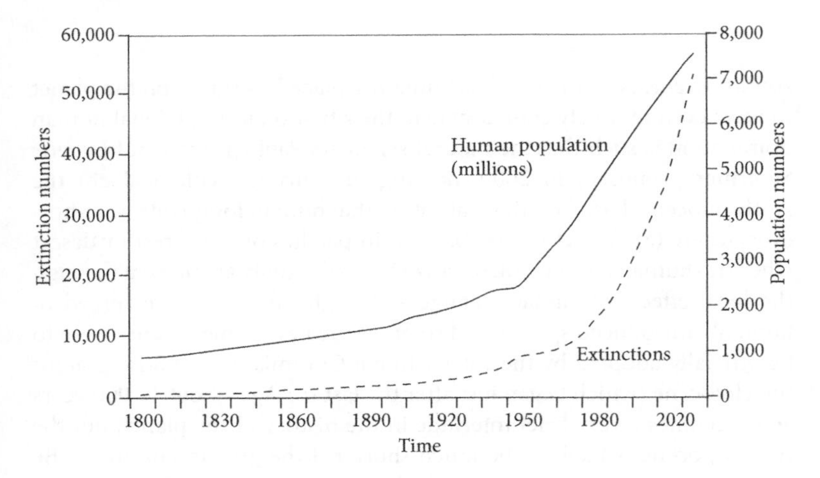

Figure 2.1 Species extinction since 1800

Source: Center for Biological Diversity (2015).

modifying the surroundings to satisfy our desires and needs. This is why the term 'Anthropocene' is rarely used jubilantly; it is a wagging finger and a warning sign. It signifies that the growth ethos of capitalism and relentless optimism of Enlightenment thought may be nearing their end.

* * *

Neoliberalism. In describing the contemporary world, it is difficult to avoid the term 'neoliberalism' entirely. Although the word has clear ideological connotations, it can be defined accurately as a means of explaining the global economic shift leading to the widespread occurrence of overheating effects. Neoliberalism refers to a particular kind of market-oriented economic ideology and practice characteristic of the late twentieth and early twenty-first centuries. It is commonly agreed that it began in earnest with the policies of deregulation and privatisation instigated in the US and UK around 1980, under Ronald Reagan and Margaret Thatcher's respective leaderships. The structural adjustment programmes implemented by the IMF in the Global South in the 1980s and 1990s conformed to the same principles, cutting public expenditure and encouraging the development of competitive markets wherever possible. This set of policies, widely known as the Washington Consensus, was at the time the outcome of an agreement between the IMF, the World Bank and the US Treasury Department.

Geographer and social theorist David Harvey defines neoliberalism as follows:

> Neoliberalism is ... a theory of political economic practices that proposes that human well-being can best be advanced by liberating individual entrepreneurial freedoms and skills within an institutional framework characterized by strong private property rights, free markets, and free trade. The role of the state is to create and preserve an institutional framework appropriate to such practices. (Harvey 2005: 2)

Neoliberal policies have since the 1980s been pursued, to varying degrees, by governments in most parts of the world, fully or partly privatising formerly public enterprises, such as railways and postal services, and encouraging, at least in theory, an unfettered market economy (although restrictions are usually placed on imports in the form of tariffs, and key industries are often heavily subsidised). The origin of neoliberalist thought is generally traced to Friedrich Hayek and his successors, notably Milton Friedman, whose finest moment may have been in the early 1980s, with the simultaneous implementation of this economic ideology in the US and the UK. However, there is an immediate precursor to these 'freshwater economists' and their critics, which is a reminder that criticism of marketisation is not new, even if global economic integration is. Hayek's teacher in Vienna, Ludwig von Mises, was an enthusiastic libertarian, an enemy of socialism in all its forms, and a believer in deregulation of markets. Mises' most important critic was the economic historian Karl Polanyi (1957 [1944]), whose *The Great Transformation* almost immediately caught the attention of anthropologists upon its publication in 1944. This book was actually the main source of inspiration for the subsequent 'great debate' in economic anthropology between substantivists and formalists, a debate which continues to this day, in new guises, across the field of economic anthropology (Hann and Hart 2011). The question concerned whether the economy should be viewed mainly as maximising behaviour (formalism) or as the social organisation of production, distribution and consumption. The conflict between marketisation and the 'human economy' represented in Polanyi's legacy can be seen as an ideological version of this 'great debate': should the aim of an economy be to generate growth and profits, or should it mainly satisfy human needs?

The Great Transformation begins on a dramatic note with the author stating, as a matter of fact, that '[n]ineteenth-century civilization has collapsed'. What he has in mind is the ultimate outcome of nineteenth-

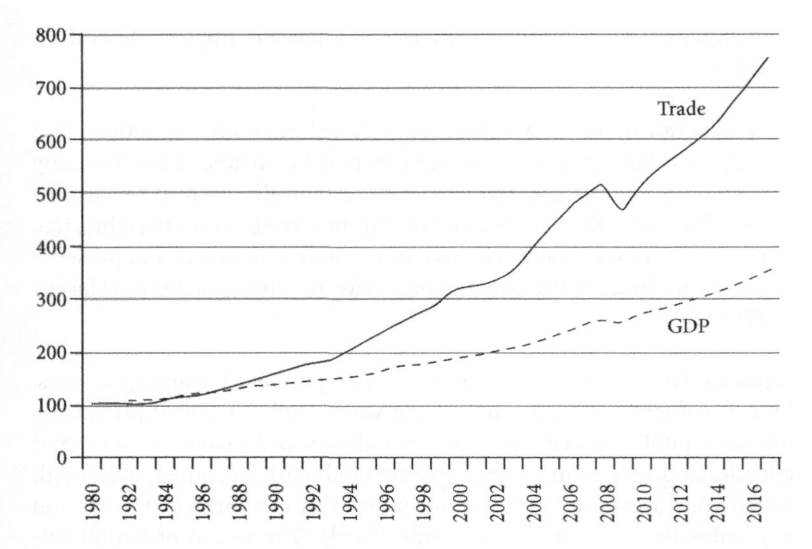

Figure 2.2 World GDP and global trade since 1980

Source: IMF World Economic Outlook (IMF 2016).

century industrialisation and colonialism, whereby the market principle became predominant and pervasive in Western societies. Polanyi uses the term 'disembedding' to describe the abstract character of capitalist market economies, where 'the economy' is accorded a life of its own, independently of actual social relations (see also Eriksen 2014a). In what is virtually an *avant la lettre* criticism of neoliberalism, Polanyi argues that the values and practices of sociality, based on reciprocity and solidarity, are more fundamental to human existence than the disembedding and ultimately dehumanising market principle. He predicts that they will prevail in the long term. A non-Marxist socialist, Polanyi argued against the commodification of labour and, more generally, the limited vision of mainstream economics. His main target was Mises, an ideological father of neoliberalism. Polanyi was not opposed to market principle as such, and was well aware of the existence of useful, well-functioning markets in non-capitalist societies. What he objected to was its spread into social domains which should be governed by principles of sociality. Just as *Gemeinschaft* is ontologically prior to *Gesellschaft* in Tönnies' classic analysis of the transition to urban, industrial society, a 'human economy' based on reciprocity and redistribution (Hart et al. 2010) is fundamental to social life, and people living in communities everywhere will therefore resist market dominance.

Polanyi predicted that unfettered market liberalism, which had been tried before the First World War, would be defeated by the principles of reciprocity and redistribution. During the first decades after the Second World War, it seemed as if he had been proven right; however, the market principle as the sole or main regulator of human economic activity would return with a vengeance in the 1980s, this time in an increasingly globally interconnected world. Incidentally, as in Polanyi's time, the so-called free markets remain dependent on steady government interventions for their functioning.

Characteristic of the neoliberal world order is the global reach of communication, but also the lack of a single, powerful, ideological alternative. The development paradigm has exhausted itself. State socialism has gone out of fashion almost everywhere. Disillusion with large-scale schemes to improve the lot of humanity is chronic, as shown by James Scott (1998) in his magisterial *Seeing Like A State*, which describes the way in which large-scale schemes devised by twentieth-century states fail owing to the mismatch (or clash of scales) between the level of abstract planning and local circumstances. However, alternatives do exist, even if they are often unarticulated or only partly articulated, but they are locally embedded, sprawling and diverse. The very diversity of the groups engaged in alterglobalisation, from peasants to unionists, from indigenous groups to industrial workers, indicates that the complaints are universal and global, while the solutions are particular and local.

* * *

Runaway processes. The anthropologist and system theorist Gregory Bateson (1904–80) was perhaps one of the most under-rated thinkers of the twentieth century. Although he is far from forgotten, he does not have an established place in the Western pantheon of great minds. Perhaps he was too restless, moving from field to field; perhaps his writings were too unsystematic and scattered (it is tempting to think of him as a Socrates in need of his Plato), or perhaps his ideas were simply too radical to be properly understood. A fundamentally ecological thinker, Bateson saw the world as consisting of relationships and processes. He also tried to show that systems which were otherwise very different might have some of the same properties (Bateson 1972, 1978). His term 'schismogenesis', for example, refers to what I call runaway processes in this book (cf. Deacon 2012), namely mutually reinforcing growth processes which eventually lead to collapse unless, as Bateson points out, a 'third instance' enters into the process and changes the relationship. He saw schismogenesis in phenomena as different as marital conflicts and armaments races.

One way of describing overheating is to see it as the confluence of several runaway processes, forms of growth that were meaningful and purposive for a long time before reaching a point where the unintended side-effects were threatening to become more noticeable than the intentional or functional effects. A famous study of a runaway process in anthropology is Roy Rappaport's *Pigs for the Ancestors* (1968), an analysis of 'pig cycles' in a Melanesian society, where the number of pigs eventually reaches a point where they become a crop-destroying nuisance rather than an economic asset. In an overheating world, runaway processes of this kind are ubiquitous.

The big question, however, is whether the runaway processes we are now witnessing, which reinforce each other and operate without any other thermostat than the finiteness of the world's physical resources, can be likened to a threatening Irish elk or just to a harmless peacock's tail.

The Irish elk, strictly speaking a giant deer, lived not only in Ireland, but in large parts of northern Eurasia until after the last Ice Age. Adult stags were more than 2 metres tall, and their antlers could weigh as much as 40 kilos. The antlers had gradually grown, partly as a result of sexual selection – those with the largest antlers attracted more females – but also as a consequence of the body size. When the Irish elk became extinct, it may have been as a result of its antler size. A fully grown set of antlers contained 2.1 kilos of nitrogen, 7.6 kilos of calcium and 3.8 kilos of phosphorus (Moen et al. 1999). It had to eat 3.8 tonnes of plants a year just to satisfy its need for phosphorus. Thus, the Irish elk may have died out as a result of malnutrition. Or it could have been driven to extinction by climate change. When the climate in Europe grew warmer and forests replaced grasslands, it would have been difficult for the large stags to manoeuvre through their surroundings, compared to smaller and more flexible herbivores, not to speak of antlerless carnivores. So the reproductive advantage of large antlers clashed with the disadvantages, or unintended side-effects, of nutritional and mobility requirements.

The spectacular male peacock's tail is also a product of sexual selection, as a large and colourful tail attracts females. However, in spite of the obvious disadvantages of dragging a long, heavy tail around, not least in confrontations with predators, the peacock has survived in the wild up to now. Its tail is striking and spectacular, a source of pride and vanity – useless, but ultimately harmless, unlike the antlers of the Irish elk, which ended up being deadly. Both are instances of runaway selection, however, locking their bearers into competitive spirals with their fellow creatures and reducing their flexibility.

The underlying question, in a neoliberal, globalised, overheated world trapped in the double bind between growth and sustainability, is, as one might guess, whether contemporary world civilisation resembles a giant elk or a peacock the most: is it patently self-destructive, or is it just unnecessarily flamboyant but mostly harmless?

* * *

Treadmill syndromes. At the micro-level, runaway processes often take on the character of treadmill syndromes, as both of the above examples from the animal kingdom suggest. These forms of competition are known in evolutionary biology as 'red queen phenomena', in homage to Lewis Carroll's *Through the Looking-Glass*, where this curious aspect of competition in natural selection was first described (Carroll 2003 [1872]; see also Van Valen 1973; Ridley 1993). When Alice meets the terrifying Red Queen, she is told to run as fast as she can. She does, but discovers that although she has run to the best of her abilities, she remains stuck in the same place. The Red Queen shrugs and confirms that in her country, you have to run as fast as you can, just to stay in the same place. In our 2012 book, the biologist Dag O. Hessen and I explored this aspect of competition in a wide range of areas, from biological evolution to technological change, advertising, sport and academic publishing (Hessen and Eriksen 2012). Since your competitors improve, or the environment changes, you have to improve and adapt merely to keep your place in the ecosystem, in a market, or in an academic hierarchy. Although we saw intensified treadmill competition in today's world, some of it destructive both in environmental and existential ways, we did not then connect it to overheating or accelerated global change. However, treadmill competition is simultaneously a premise for, an integral part of and an outcome of the runaway processes that create an overheated world.

* * *

Double bind. A second concept from Gregory Bateson's (1972) ecological thinking is that of the double bind. He originally developed it with colleagues in the context of a theory of schizophrenia which emphasised pathological family relations and communication rather than physiological factors. A double bind is a self-refuting kind of communication, as when you say two incompatible things at once. A person trying to act on the basis of a double bind will never be able to do it right, since no matter what they do, it can be objected to. In the contemporary world, the world of the Anthropocene and neoliberal runaway growth, the double bind of growth and sustainability is a fundamental contradiction. It seems impossible to have it both ways. In the daily life of a member of the global middle class, the pressure to behave in an environmentally responsible way may be strong, and they recycle, use public transport rather than driving, try to buy organic, locally produced food and so on. At the same time, they may occasionally travel by plane for business or pleasure, and their lives depend completely on a carbon economy. Similarly, at a higher level of scale, in many parts of the world, business leaders and politicians have begun to talk about sustainability and climate politics, while simultaneously favouring economic growth, which nearly always implies increased energy consumption.

A double bind is more fundamental than a dilemma. In the political realm, a dilemma increasingly relevant for social reformers and critics is that emerging in the encounter between class politics and green politics. Class politics is usually based on demands for improved material conditions for workers, while green politics favours a change in the economy and a *de facto* reduced material standard of living. Pragmatic solutions to this dilemma have nevertheless been proposed: social justice and equality can, in theory, be reconciled with a high quality of life and green politics. This would not be the case with a true double bind, such as that between fossil growth and sustainability.

* * *

Flexibility. The third Batesonian concept I draw inspiration from is flexibility. The term has an everyday meaning, but it can also be defined precisely. In 'Ecology and flexibility in urban civilization' (in Bateson 1972), Bateson defines flexibility as 'uncommitted potential for change' (1972: 497). The context to which he related was the emerging environmental degradation which first attracted widespread attention in the early 1970s. Bateson argued that increased energy use entails a loss of flexibility in the sense that it shrinks the opportunity space. In a society which is built around the daily use of the car, for example, it is difficult to revert to slower and less energy-intensive means of transportation. Shops, services and workplaces may only be accessible by car. More fundamentally, his view was that the flexibility used (and used up) by growing populations harnessing much of the available energy for their own purposes, reduced the flexibility of the environment. This perspective, naturally, is important on an overheated, 'full' planet.

Bateson describes a healthy system, flexibility-wise, by drawing an analogy with an acrobat on a high-wire.

> To maintain the ongoing truth of his basic premise ('I am on the wire'), he must be free to move from one position of instability to another, *i.e.*, certain variables such as the position of his arms and the rate of movement of his arms must have great flexibility, which he uses to maintain the stability of other more fundamental and general characteristics. (Bateson 1972: 498)

Maintaining flexibility in the system as a whole, Bateson adds, 'depends upon keeping many of its variables in the middle of their tolerable limits' (1972: 502). In order to use the term accurately, it is thus necessary to specify the parameters limiting the upper and lower threshold values, and also to demonstrate the significance of wider systemic connectedness which affects, and is affected by, flexibility in the realm under investigation.

In cognitive theory, a major theoretical issue concerns the way in which knowledge is being selected, sifted and organised. In a critical overview of the state of the art at the turn of the millennium, Tor Nørretranders, in *The User Illusion* (1999), distinguishes between *information* and *exformation*, the latter being potential information that is consciously or unconsciously selected away or filtered out. Dan Sperber and Deirdre Wilson's (1986) notion of *relevance*, informed by linguistics and Darwinian selectionism, is a kindred concept. While the dynamics of information exchange and knowledge development are not random, they are also far from predictable. There is much, much more potential knowledge present in our surroundings, in our brains and in our bodies than what is being used. The 'uncommitted potential for change' is very considerable.

In one of the late twentieth century's controversies about natural selection, Stephen Jay Gould and Elizabeth Vrba (1981; see also Gould 2002) introduced the term 'exaptation' to denote phenotypical features whose functioning had undergone change due to changes in the wider system. These structures were, in other words, flexible and responded, like the acrobat on the high-wire, to changing parameters in their surroundings. The human hand is a uniquely flexible structure; like good basic research (as opposed to applied research), it has no specialised application, but it can be used for lots of different things.

In another debate about natural selection, Steven Rose (1996) and others argued that there cannot be a simple relationship between DNA and the organism, between genotype and phenotype, since there are important phenotypical effects arising from the interaction between hereditary material and its surroundings. Already at the level of cell chemistry, this kind of flexibility in hereditary material is evident. For one thing, it is well known that there are common characteristics in humans that are inborn but not genetic, which are caused by the mother's diet during pregnancy.

In a flexible system, many things can be done differently; it has not locked itself to one particular course. The economic practices of a free peasant on his own land, who may have a few animals of different species and grow a dozen different crops, are accordingly more flexible than those of an industrial farmer specialising in soybeans or cattle. Standardisation and specialisation reduce flexibility. At the same time, increased flexibility in a certain area often leads to reduced flexibility in others. With mobile information technology, work becomes more flexible in the sense that you can do your job anywhere; but you only become more flexible as regards place. You become less flexible regarding time, since much of your elbowroom is used up through constant availability. Also, one man's flexibility may entail another man's imprisonment.

Importantly, at a planetary level of scale, the increased flexibility resulting from the fossil fuel revolution – the freedom to move swiftly and to consume, the electrification of homes and so on – has dramatically reduced the flexibility of the global system. With a growing global population of over 7 billion, reverting to a pre-fossil fuel economy may be necessary, but it is difficult to imagine in practice. The Green Revolution in India enabled the country to leave famines behind, but the country seems to have painted itself into a corner. The population has grown massively since the introduction of improved techniques, better fertilisers and new cereal strains in the 1970s, which also means that the country has now committed itself to particular crops and techniques, and even if the long-term side-effects turned out to be detrimental to the system (for example impoverishment of the soil), India cannot return to the pre-Green Revolution agricultural practices. There has been a loss of flexibility in this domain, which can only be regained through innovations enabling simultaneous high food production and long-term ecological sustainability.

Growing flexibility in one field tends to lead to the loss of flexibility in another. However, the model does not necessarily result in a zero-sum game: in some cases, conditions of 'matching flexibility' are achieved, that is to say complex systems where a desired level of flexibility is maintained in both, or all, of the relevant interacting subsystems. When there is no attention to matching flexibility, the relationship between subsystems becomes skewed. Bateson's main example is the relationship between civilisation and environment, as he puts it. Modern civilisation becomes ever more flexible in terms of cultural production, individual choice and so on, and as a result the culture–environment relationship loses flexibility because of increased dependency on massive exploitation of available energy and other resources.

* * *

Reproduction. A close synonym of sustainability, reproduction refers to the ability of a person, a system or a social field to continue on its path without constantly having to adjust to exogenous changes. Much of the opposition to the runaway processes of globalised overheating can be understood as *crises of reproduction*, that is, ruptures in the system or life-world resulting from accelerated, imposed change in one or several crucial realms. Globalised crises in the economy, the environment and identity are being experienced, and dealt with, almost everywhere in the world. Since social reality is created through the interaction of individuals, networks and communities and their relationships with their wider environments, the crises as such are bound to differ from place to place. Put differently: globalisation does not create global persons, and the standardising, large-scale changes imposed through the neoliberal world economy and climate change do not have the same results everywhere, simply because locations remain different and, in important respects, unique. Mining takes place in both Queensland and in upland Sierra Leone, but it is not the same thing in these two locations. Glaciers melt in the Andes and in Greenland, but local responses are strikingly different. Financial bubbles burst and lead to worker precarity from South Korea to Greece, but local understandings and reactions differ. In many Muslim countries, it is a fairly common assumption that the recurrent global financial crises are more or less deliberately created by the Americans and Israelis to strengthen their grip on the world economy, and YouTube videos supporting this view can easily be found; understandings of global economic crises can be very different elsewhere.

Faced with the perceptible and sometimes dramatic impact of local events which have their origin in distant lands or at a staggering level of abstraction, people everywhere experience problems of reproduction: economically, culturally and environmentally, they see their autonomy and their right to define themselves and their destinies as being threatened. They are confronted with their own vulnerability, they begin to doubt who – or what – they can trust, and who – or what – they can blame; and they develop a heightened awareness of risk. Whether they adapt and adjust, protest or delink is an empirical question; and this is why it is necessary to look at the crises of globalisation from the vantage-point of the crises of reproduction at the local level.

* * *

Scale. Global crises are rarely perceived locally as such. Their local repercussions or more immediate effects are perceived, rather, as crises of reproduction and a threat to or loss of autonomy. The typical orientation, even in a globally integrated world, is small scale, although we oscillate cognitively between levels of scale ranging from intimate relationships to the entire planet. I distinguish between four kinds of scale, each of which can be observed at every level, from the planet as a whole to the grain of sand on the beach.

In a very general sense, scale simply refers to the scope and compass of a phenomenon – whether it is small or big, short-term or long-term, local or global. We may nevertheless be more specific. Scale can be taken to refer to a combination of size and complexity (Grønhaug 1978). Social scale can be defined as the total number of statuses, or roles, necessary to reproduce a system, subsystem, field or activity; in other words, if two societies have 10,000 inhabitants, they are of differing scale if their respective divisions of labour vary significantly. Large-scale systems depend on the contributions of many persons and require an infrastructure capable of coordinating their actions, monitoring them and offering a minimum of benefits enabling persons around the system to reproduce it. In this sense, scale is a feature of social organisation.

In a different, but complementary sense, scale can refer to cultural and individual representations of society, the world or the cosmos, and there is no necessary congruence between social scale and cultural scale. A society may be embedded in global networks of production and consumption without its residents being aware of their place in a global system. Conversely, residents of societies which are relatively isolated in terms of economic and political processes may be well connected through symbolic communication and possess a high awareness of their place in wider, global systems. Fredrik Barth once compared two societies, in which he had done fieldwork, along such lines. The Baktaman of Papua New Guinea lived in small social and cognitive worlds, while the Basseri nomads of Iran were embedded in relatively small-scale economic systems and had a small-scale social organisation with a limited division of labour, but at the same time, they had a deep awareness of Persian history, recited classic poets and asked Barth questions about sputniks and the armaments race (Barth 1961, 1975; Eriksen 2015).

Finally, temporal scale is important, not least in the context of environmentalism and industry. Environmentalists often assume that industrial capitalism is short-sighted, while only ecological thought takes the long, planetary perspective. However, mining in Australia (and elsewhere) presupposes long-term investment which is sometimes expected to yield profits only decades ahead; and, conversely, it is difficult to document large-scale environmentalist movements where actors take decisions solely based on assumed consequences that will only become apparent years after their own demise. The time scale on which people take decisions is relevant in a comparable way to the cultural scale by which they orient themselves, and the social scale in which they are integrated through networks and social organisation.

In a word, I distinguish between *social* scale (the reach of your networks), *physical* scale (the compass of an infrastructural system), *cognitive* scale (the size of your perceived world) and *temporal* scale (the time horizon you imagine, forwards and backwards, when taking decisions and making plans). Solutions to global crises typically include pleas to scale up socially and spatially (world government, more power to the UN, international climate agreements and so on) and equally passionate pleas to scale down socially and spatially (small is beautiful, local economies are more equitable and sustainable, and so on), while simultaneously scaling up cognitively and temporally (act locally, think globally).

In general, the large-scale phenomenon is standardised, while the small-scale phenomenon is unique. Both are for better and for worse. A *clash of scales* occurs when the intersection of two or more levels of scale leads to a contradiction, a conflict or friction. For example, most policies are decided at the local or national level, whereas climate change is a global problem, and all countries are interlinked through international trade, mobility and communication networks. At the same time, since political decisions are taken at the state or even transnational level, a result can be local resistance owing to a sense of alienation and disenfranchisement. At the cognitive level, the knowledge systems underpinning policy are based on abstract scientific methods, which may contradict knowledge or experience at the local level. It is also a typical outcome of globalisation that the 'economies of scale' favouring large-scale operations make formerly viable, small-scale activities unprofitable. The time scale is also important, and there is a clash of scale in this dimension when policy makers are forced to consider the next election whereas emissions may have very long-term impacts. Time scales concern individual and collective horizons as well, and could be a person's remaining lifetime, that of her children or grandchildren, future generations or the planet as such. In this way, temporal scales are consequential for the ways in which climate change prevention/mitigation efforts are implemented. Whenever someone makes predictions about 'the future', it is always useful to ask when this future is expected to happen.

* * *

Neoliberalism and information age global capitalism have affected communities everywhere and are simultaneously both universal and global, and locally particular. People will always understand themselves in terms of their enduring social relations, their webs of reciprocity and moral obligation, their shared intimacies and structures of interpersonal trust. This is why disembedding processes producing an ever more abstract world, whether driven by states, markets, corporations or NGOs, are bound to be partial, and will be met with resistance when imposing changes that threaten people's autonomy and integrity.

As I have argued, it was in the 1990s that modernity shifted to a higher gear, and acceleration has continued since the turn of the millennium. It has long been predicted that China would play a leading role in the world economy, but it is happening only now, in the 2010s, and the global importance of China remains huge, despite its growth slowing recently. The so-called social media, which used to be called Web 2.0 around 2005, that is, the internet platforms and services based on communication rather than information, have flourished only since the mid noughts. This is also the case with sophisticated methods of mass electronic surveillance and forms of social control relying on 'big data'.

Indicators continue to point steeply upwards in innumerable areas of human activity, transforming the conditions of life (human as well as non-human) as I write. Just after the First World War, the sociologist William Ogburn proposed the term 'cultural lag' (Ogburn 1922) to describe a situation where ideas and concepts about the world lag behind changes in the physical world. A non-Marxist variety of the relationship between infrastructure and superstructure, Ogburn's concept calls attention to the tendency to use old ideas to understand new phenomena. Perhaps we are still, to some extent, using explanatory models from the early twentieth century in attempting to understand the early twenty-first century. The world is increasingly difficult to understand; it changes fast and, at such a time, cultural conservatism is likely to have a strong presence. Fundamentalists of all kinds, from biological reductionists to religious fanatics, are capable of mobilising huge armies of grateful supporters. When there is movement all around, stability becomes a coveted resource.

The post-1991 world is an overheated world, above all characterised by frictions and tensions. The networks connecting us to each other are denser, faster and richer in consequences than ever before. The growth of urban slums throughout the Global South is an indirect result of economic globalisation (Davis 2006), as is the growth of a transnational precariat (Standing 2011), a mobile labour force deprived of rights and predictability. The growth of militant Islamic identity politics across the Muslim world, the rise of a motley group of new social movements opposing global neoliberalism (Maeckelbergh 2009), the dramatic increase in social inequality in the US, and the double bind between economic growth and environmental responsibility all indicate that globalisation is not a linear process entailing increased enlightenment and prosperity. Yet the networked capitalist world is a framework, or scaffolding for almost any serious inquiry into cultural and social dynamics today, not least those movements that reject the hegemony of neoliberal capitalism; labelled 'anti-globalisers' at the outset, they prefer the term 'alterglobalisers', as it is not interconnectedness as such they oppose but its neoliberal form. The transnational Islamist movement, the green movement and the new social movements try to show, in various ways, that a different world is possible, while simultaneously pointing out that accelerated change is a reality, and that globalisation is fraught with structural contradictions and local conflicts. As any traffic researcher will confirm, speed requires space, and traffic is becoming more dense in the transnational highways. Nor is there always agreement about the traffic rules.

This is an accelerated world, where everything from communication to warfare and industrial production takes place faster and more comprehensively than ever before. Speed, in physics, is closely related to heat. In other words, when you say of someone that he or she is suffering from burnout, the metaphor is an apt one. The burnout is a direct consequence of too much speed. But the heat metaphor is also appropriate in other areas. When there is a crash in the stock exchange, financial brokers speak about a 'meltdown', and when rates increase too fast, they talk about the need to 'cool down' the economy. Riots and violent demonstrations are often associated with overheated tempers. Finally, climate change is connected to overheating in two main ways: world temperature is increasing, and the cause is accelerated change, especially with regard to energy use.

Perhaps this is a reason why global warming has rapidly become a key narrative about our present and near future. It fits with the other stories about the present that are being narrated. By focusing on heat as an unintended consequence of modernity, the concept of global warming fits perfectly with other stories about the present era. It can almost be seen as a natural-science version of the stories about ethnic, religious and cultural frictions, runaway urbanisation, large-scale transformations of nature and technological development which has got out of hand. They are all about a world which is *trop plein*, filled to the brink and on the verge of spilling over.

Moreover, the different forms of accelerated change interact and magnify each other through complex feedback loops. Speed is contagious, and friction in one area can often lead to friction elsewhere. A volatile economic situation, marked by rapid changes which affect different people in different ways, leads to defensive reactions among those who are not among the beneficiaries; the fast growth currently witnessed in extractive industries, from fracking to coal and iron ore, leads to local conflicts as well as climate change; the world of instantaneous communication and global networks simultaneously enables state surveillance and the fast spread of political ideas, ranging from conspiracy theories to religious movements; and fast migration has led to political polarisation and overheating effects among natives as well as immigrants in the realm of politicised identity. It is a new world, and it has to be studied from below as well as from above.

3. Energy

Dystopian science fiction books are always, or nearly always, about the social life of energy. Scenarios tend to hover between a focus on the squalid life afforded to what is left of the human species after the exhaustion of abundant energy supplies and the dismal environmental destruction caused by greed, growth spirals and a lack of foresight. Abundant energy, the very lifeblood of complex societies, enters into human lives in a multitude of ways, and owing to scarcities, inequalities and environmental concerns, thinking and acting with energy now takes on a new urgency, and the issues are being addressed in diverse and imaginative ways. Some experts believe that solar power will save the global climate and ensure continued economic growth without severe side-effects (but even solar panels require rare, scarce minerals). Others emphasise the continued importance of coal, gas and oil, pointing to the impossibility of prosperity and material security in the Global South without a large supply of fossil fuels. Yet others look at a possible shift towards renewable sources of energy as an efficient way of decentralising power, since energy consumers will be capable of producing their own energy. Finally, human desires, hopes and dreams often clash with demands for sustainability. This is the case both in the energy-abundant and the energy-deprived parts of the world (Winther 2008; Wilhite 2012). In an overheated world, it can be difficult to reconcile global sustainability with common dreams and notions of the good life, whether they are formulated in an Indian city, a Pacific island or an African village. As Trawick and Hornborg (2015) have argued, the cost of increased energy use is necessarily environmental decline. Their conclusion, in an article about thermodynamics and the economists' fiction of unlimited good, is that 'economic "growth" is a physically destructive process rather than a creative one, based on the consumption of limited stocks of natural resources that are being rapidly drawn down and made more scarce every day at a planetary level' (Trawick and Hornborg 2015: 16). With this claim, they echo the view from the *Limits to Growth* report (Meadows et al. 1972), where the central argument was that a growing world population committed to industrialism would run out of crucial resources in a foreseeable future. Their predictions, which have largely been vindicated by later studies (see for example Turner 2014), were all the more impressive for having been formulated before global warming

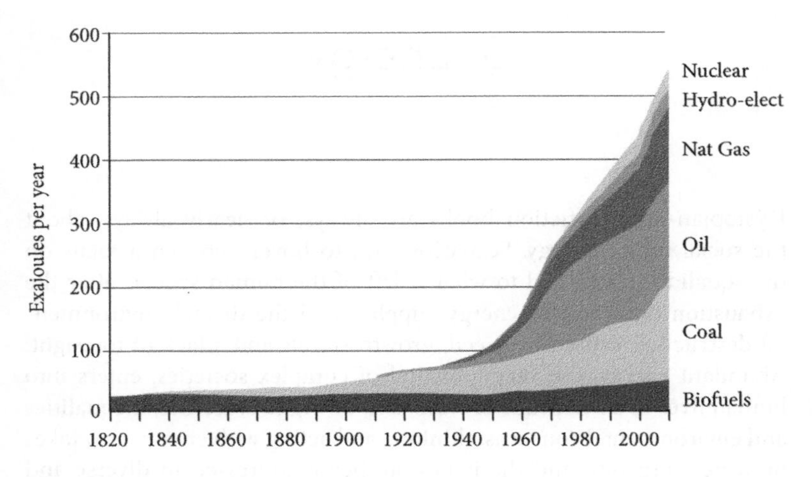

Figure 3.1 World energy consumption since 1820

Source: Our Finite World (2015).

was on the scientific agenda, and at a time – a generation ago – when world population was about half of its present size.

Coal planet

On the map, Gladstone looks tempting enough to a frozen northerner. The city is located just outside the Tropic of Capricorn on Australia's scenic eastern coast, with a perfect climate, a scattering of subtropical islands just off the coast and the southern fringe of the Great Barrier Reef a few miles further out. But if you go there, what you encounter is an industrial city literally marinated in coal and gas. The harbour area is dominated by two large coal terminals and massive heaps of the black stuff. The railway tracks do not transport happy tourists, but black, bituminous coal from the Bowen basin in the Queensland outback. The seemingly interminable coal trains dump their cargo in the docks, where it is watered to prevent spontaneous combustion.

In Gladstone, even the sunset is sponsored by the mining industry. Since the sun sets in the west, it is impossible to watch its daily descent without simultaneously looking at three slim, tall, symmetrical columns to the right. They are the chimneys of the coal-fired power plant, the largest in Queensland and the pride of Gladstone.

Drag your finger along a rail or a piece of garden furniture, and you come to the realisation that Gladstone is coal. When the breeze blows from the Pacific, a fine dust settles everywhere.

It is a proud high modernity, brimming with self-esteem and skilled information officers that you are likely to meet in Gladstone. The air also smells surprisingly clean. The coal burnt in the city is generally free of smell in rich countries such as this, since it is mostly converted to electricity. This was not the case a few decades ago or so, certainly not in Eastern Europe. You noticed the stink the moment you walked out of the U-Bahn in East Berlin. The air was acid, toxic, sour. It smelled of humourless, grotty totalitarianism. It was the tang of low-grade East European lignite (brown coal). But the same smell, or a similar one, had until recently predominated in many cities in the West as well, until they switched to electricity for heating, London being one of the last. Even when thousands died following the 'Great Smog' in 1952, Londoners did not cease to burn coal, in private homes as well as factories and offices. For what would the alternative have been? In *1952*? The smoke rising from the furnaces was poisonous and smelly, but it also smelled of prosperity and development. One wonders if this is also a common notion in Indian and Chinese industrial estates today, where coal is still a source of smog, and where respiratory ailments are far more widespread than in the countryside?

So how did they produce all that electricity which eventually took over from coal? With coal, naturally, what else? Coal and modernity are two sides of the same coin; coal and the post-Napoleonic accelerated change are Siamese twins. Without coal, no industry; without coal, no exponential population growth, no steamships, locomotives or factories. Without coal, the English Industrial Revolution would have ground to a halt after just a few years. By then, there would have been no trees left on the British Isles.

It is no coincidence that the earliest modern industry, with a few exceptions, was created near large coal deposits which were easily accessible: the Midlands, South Wales, Belgium, northern France, the Ruhr Valley, Pennsylvania. In the early nineteenth century, the coal was used, among other things, to produce steel for railways and locomotives, and to get the trains moving. By mid century, a network of railways and steamships enabled industry to spread to a larger area, since the coal could now be transported inexpensively and efficiently.

Coal had already been used for several thousand years. It exists in many forms of varying quality. Charcoal is the simplest, anthracite the most complex, energy-rich and compact form. Where mineral coal deposits lay near the surface, they were used as an energy source on a small scale in agricultural societies. Charcoal was used in the smithies.

It was the enormous optimism and positive energy released by the European scientific revolution and the Enlightenment – all of a sudden, humans felt up to playing God – that created the technology which enabled miners to dig ever deeper to get at the coal and use it for ever more purposes. Mining shafts were secured with beams, steam-powered pumps removed excess water from the tunnels, and already a couple of hundred years ago, the brand-new factories of Manchester could utilise coal which had been dug out several hundred feet below the surface. The marriage of the steam engine and fossil fuels was the killer app of the nineteenth century and, even now, British nostalgia for the age of coal and steel is evident in museums, in popular literature, in lovingly restored station buildings and steam engines.

By this time, we were engaged in earnest with the transformation of the planet, and the speed would continue to increase. The exact dating of the onset of the Anthropocene remains open to discussion, but a meaningful starting point would be around the year 1800, the beginning of the fossil fuel age. Whereas religion had given hope and comfort to people for thousands of years, the story of progress, development and promises of a better life before death would now take over as a source of hope and expectations, certainly in Europe, where religion was slowly losing its authority. The story of hope as progress is intimately connected to increased energy use, and now that it is losing its general appeal, it is no coincidence that this is happening just as people begin to realise that the unintended side-effects of coal and its relatives are serious and dismal.

It took *Homo sapiens* 200,000 years to reach the first billion. Then, coal entered the world of humans on a large scale. Since then, we have fulfilled Jehovah's command exceedingly well: 'Be fruitful and increase in number; fill the earth and subdue it. Rule over the fish in the sea and the birds in the sky and over every living creature that moves on the ground' (Genesis 1:28). But keep in mind that while population has grown sevenfold since 1800, energy use has grown 28 times. The graph points optimistically upwards, and at the moment, there is no indication that it will soon flatten out, unlike population growth, which is slowing down in most parts of the world (Africa being the main exception).

Coal is a wonderful source of energy; it is compact, easy to load and transport, and it withstands both warm and cold temperatures pretty well without deteriorating. Coupled with the appropriate machinery, coal increases productivity many times, both in industry and agriculture; it liberates millions from physical work and makes it possible to feed growing and increasingly urban populations.

Coal is condensed sunlight. It has taken the planet up to 300 million years to build up its stores of this wonderful energy source, once described by Jean-Paul Sartre as the gift of other living beings to humanity – and we are now burning it up in a matter of a few short human generations, and at the same time destabilising the climate. Obviously, researchers and commentators say, this cannot continue indefinitely but, as with other forms of addiction, quitting is hard, especially when you are trapped in a double bind.

Some may have got the impression that the time of coal is more or less over; that it has by and large been replaced by other fossil fuels. After all, we no longer smell the smoke and breathe the dust in the West. And it is clearly true that there are some advantages to oil and gas, above all that their production and distribution do not require as much labour input as coal. Coal is extracted by men and transported on rails, while oil and gas are extracted with pumps and transported through pipelines. The risk of striking workers is much reduced in this way. Had Saudi Arabia been rich in coal rather than oil, it is less likely that the country could have continued to exist as one of the most totalitarian states in the world, managed as the private property of the royal house of Saud under the protection of the US. With coal mines instead of oil wells, it is conceivable that hundreds of thousands of Arab proletarians would have discovered socialism and formed trade unions, as they did elsewhere. As Timothy Mitchell writes in *Carbon Democracy* (2011), political power is not just related to the ability to produce, control and distribute energy, but also the ability to stop the flow of energy. For this reason, coalmines are more democratic than oil platforms. In mining communities, the people have a real opportunity to seize power; in oil societies, from Nigeria to Turkmenistan, power tends to be concentrated and centralised.

> Between the 1880s and the interwar decades, workers in the industrialised countries of Europe and North America used their new powers over energy flows to acquire or extend the right to vote and, more importantly, the right to form labour unions, to create political organisations, and to take collective action including strikes. (Mitchell 2011: 26)

In a word, the workers' rights that were achieved during the first century of industrialism in the West, and which many now see evaporating in a more fragmented information economy (for example Sennett 1998), were largely a result of the workers' *de facto* control over energy flows.

Political elites have long been aware of the mobilising potential among energy producers. Mitchell reveals that as 'early as the 1940s, the architects of the Marshall Plan in Washington argued for subsidising the cost of importing oil to Western Europe from the Middle East, in order to weaken the coal miners and defeat the left' (Mitchell 2011: 236). This was due to the awareness, following a century of experience with a radicalised coal miner class, that '[t]he coordinated acts of interrupting, slowing down or diverting its movement created a decisive political machinery, a new form of collective capability built out of coalmines, railways, power stations, and their operators' (Mitchell 2011: 27). The shift from coal to oil meant reduced bargaining power on the part of the working class. Interestingly, contemporary coal mining has been restructured and reconfigured to resemble oil drilling formally. The typical Australian coal mine, for example, is not a network of underground tunnels connected to mining shafts, manned by hundreds of dusty, resentful workers pushing heavy carts on rusty rails. It is a Martian landscape where vegetation and topsoil have been brutally removed with huge machines in order to get to the coal more easily. The typical Australian coal miner is a well-paid white man who spends much of his working time operating a very large excavator, sitting in an air-conditioned cabin with a Pepsi Max and a headset while he digs out coal and rubble with a hydraulic shovel. Like oil drilling, this technique in coal mining is less labour-intensive and more capital-intensive than in the past, and strikes are rare.

Coal and the politics of the industrial age are closely connected. Retrospectively, it is easy to see that the old left in Great Britain expired during the lengthy mining strike in 1984–5, when the legendary trade union leader Arthur Scargill finally lost to Margaret Thatcher's neoliberalism. Following the strike, some of the oldest coal mines in Yorkshire were closed down. Had it happened today, perhaps the environmental movement would have cheered. At the time, no radicals rejoiced. Thousands of jobs were lost, and they have not returned. Cathrine Moe Thorleifsson, a postdoctoral researcher at 'Overheating', has done research on Islamophobia and the rise of UKIP in Doncaster, one of many northern English towns that lost their main source of income and identity at the time (Thorleifsson forthcoming). The end of the coal age in England signalled, belatedly, its definite end as a world leader. Since 1800, energy use in British industry had increased by 50 per cent every decade, owing exclusively to growth in the extraction of coal. This was now over.

The English coal mines were not closed because coal was by now considered an obsolete, Victorian kind of fuel. The prosaic reason was that they were no longer sufficiently profitable and unable to compete with mines elsewhere in the world. But it is also clear, as was intimated by union leaders at the time of the last mining strike, that the Conservative government was determined to break the spirit of the unions and the unruly, leftist working class in the mining areas and beyond.

Coal remains a popular source of energy. As the Figure 3.1 shows, coal is nowhere near being replaced by cleaner and more efficient forms of energy. Coal consumption continues to grow in spite of the fast growth of other fossil fuels – oil and gas – since the mid twentieth century. Between 2004 and 2014 alone, world coal production increased by 40 per cent. Indonesia, which was a nonentity in the world of coal a decade before, was by 2015 the world's largest coal exporter. That is good news for certain transnational mining companies, but not necessarily for the local populations in East Kalimantan or for the global climate.

Australia is blessed with an abundance of natural resources, including coal. More than three-quarters of the electricity in the country is produced by coal, and there is still enough left for the country to be one of the main coal exporters in the world. And Gladstone – an industrial town of *utes* (utility vehicles), coal trains, pubs and drive-through bottleshops, cranes and power stations, blokes in high-visibility clothing and blokes in short-sleeved white shirts – is a hub. In Central Queensland, all tracks lead to Gladstone. From here, the coal is shipped on to Japan, China and India. During my fieldwork there in 2013–14, I asked one of the friendly information officers in the harbour whether she thought it was such a great idea, in the long run, to produce all this coal. She quickly responded that the coal was a major source of work, value and prosperity in Queensland. And besides, she added, gazing at a large Asian cargo ship docked at the Tanna coal terminal, who were we to refuse the Chinese their Industrial Revolution?

New facilities are still being built for the storage and shipping of coal from the Queensland coast. In Gladstone itself, a huge new terminal was completed in 2015, and plans are afoot to expand coal ports further north as well. Hardly anyone in the region believes that the peak is about to be reached, that the future world will crave less fossil fuel than the present one. One of the largest industrial projects in Gladstone in the last few years has been the dredging of the western harbour in order to increase the capacity of the port: 21 million cubic metres of dredge spoil have been removed and dumped, mostly within the Great Barrier Reef region. The water was contaminated, and the marine fauna suffered. Gladstone's fishermen went out of business, or had to relocate.

When the double bind of growth and sustainability is addressed, as is increasingly the case with respect to fossil fuels, it is common to assume that the extractive industries are short-sighted and operate on a short temporal scale, while environmentalists think of long-term sustainability. As pointed out earlier, this is not necessarily the case. The coal industry presupposes long-term planning and is rarely able to generate fast profits. Coal prices declined in 2013–14, but this does not prevent the big mining corporations from expanding. Since coal mines are often commercially viable for a century, they will in all likelihood – provided markets remain hungry for coal – make money for their owners, but not necessarily next year.

Large investments reduce flexibility and, in Australia, the number of jobs, the amount of money and the extent of infrastructural investments entailed by mining is such that a transition to a carbon-neutral society is almost unthinkable. The mega-coal mines of Queensland are too big to fail. At the same time, an increasing number of Australians express anxiety – farmers, tour operators, environmentalists and locals affected by mining operations – that too much is happening too fast. Even the IEA (the International Energy Agency) now states in its annual reports that it is time to think ecologically and sustainably about energy.

History often takes surprising turns, and it stands to reason that this should happen at a time of accelerated change. Until around 2013, a flurry of books and articles took as their premise that peak oil was near, and that humanity would have to search for new energy sources very soon. For those with green inclinations, this impending energy crisis was sometimes seen as a blessing in disguise. In *Carbon Democracy*, Timothy Mitchell not only shows how the shift from coal to oil prevented a large, powerful working class from demanding rights, he also identifies the links between oil production in the Middle East and the growth of an American lifestyle totally dependent on high energy consumption. Mitchell assumes that fossil fuels will run out: 'The production of oil from conventional sources appeared to have reached, if not its peak, at least the long, uneven plateau from which it would be increasingly difficult to maintain levels of production' (Mitchell 2011: 232). In another book, which is essentially about overheating effects, namely the political scientist Thomas Homer-Dixon's *The Upside of Down* (Homer-Dixon 2006), the coming energy crunch is virtually a premise for the entire argument. As late as 2011, a major edited collection about oil and power written by anthropologists (Behrends et al. 2011) posited as a premise that peak oil had probably been reached a few years earlier. Then, with recent discoveries of new fields, the imminent availability of oil in the Arctic due to the melting sea ice, and especially the new technologies enabling affordable access to so-called unconventional fossil fuels (shale oil, coalseam gas and so on), it became more likely that the planet would face a climate catastrophe in this century than run out of fossil fuels. In Australia, the largest energy controversy in the mid 2010s concerned the Galilee Basin in the interior of Queensland, a huge coal deposit reasonably near the surface and inexpensive to extract. The influential climate activist Bill McKibben remarked, during a visit to Australia in 2013, that if the coal in the Galilee Basin were to be exploited, the 2° C temperature increase decided as a maximum in the Copenhagen Accord in 2009 (and later with the 2015 Paris Agreement) would no longer merely be unlikely, but impossible to reach.

* * *

The global dependence on fossil fuels, and the spectacular success in extracting and utilising them, represents one of the main narratives about the Anthropocene and its inherent contradictions. Hardly anyone would have imagined, when the Great Western Railway started its operations in England in 1838, that the technology behind this British success story would eventually lead to lasting environmental destruction on a global scale. The world was still pretty empty seen from a human perspective, most of it pristine. For a long time, coal and steam engines spelled prosperity, civilisation and progress. After the Russian Revolution, Lenin famously stated that socialism meant Soviets (local party chapters) and electrification.

In the current neoliberal global economy, long-term environmental consequences are definitely on the agenda, but not among those who set the terms for continued extraction. Although organisations that favour a progressive policy on climate and the resource companies both operate on comparable, transnational scales, it sometimes feels as if the IPCC (Intergovernmental Panel on Climate Change), environmental organisations and NGOs live in a parallel universe, separated by an invisible wall from any participation in actual decision making. The energy market operates on a global scale, but it is disconnected from considerations about long-term unintended consequences, which are discussed in other arenas. And although the problems of ecological sustainability are well known, the current world energy system lacks the flexibility needed to change course. In spite of considerable interest in renewables, the global economic system is locked in a path dependency on oil, gas and coal.

Thinking about energy

An anthropologist who fell out of favour decades ago for his materialist determinist tendencies – among other things, he argued that cannibalism was ultimately caused by protein deficiencies – was Marvin Harris (1927–2001). But precisely because of his unrepentant fascination with the material at the expense of the symbolically meaningful, he understood a thing or two about energy and its relationship to politics. As early as 1978, years before Edward Snowden was born, Harris wrote:

> There already exists the electronic capability for the tracking of individual behavior by centralized networks of surveillance and record-keeping computers. It is highly probable that the conversion to nuclear energy production will provide precisely those basic material conditions most appropriate for using the power of the computer to establish a new and enduring form of despotism. Only by decentralizing our basic mode of energy production – by breaking the cartels that monopolize the present system of energy production and by creating new decentralized forms of energy technology – can we restore the ecological and cultural configuration that led to the emergence of political democracy in Europe. (Harris 1978: 288)

There is no indication that humanity, all 7 billion of us (and counting), will be able to survive in a dignified and peaceful way without a steady and reliable supply of energy. At the same time, business as usual is neither likely nor defensible in the near future, given what we now know about the destructive effects of fossil fuel consumption on a planetary scale. It is difficult to deny that contemporary modernity has gone into overdrive, replete with runaway processes, and full speed ahead in many interconnected domains; and social scientists have begun to discover that the new situation offers problems not only for politicians and planners, but also for research. A considerable number of social scientists, accordingly, now carry out research on local responses to climate change. Since the publication of Susan Crate and Mark Nuttall's then pioneering *Anthropology and Climate Change* (2009), the anthropology of the environment and climate has expanded considerably (see Hastrup and Olwig 2012; Baer and Singer 2014; Barnes and Dove 2015). Looking at climate and the environment from an overheating perspective entails a focus on the relationship between capitalist expansion and environmental degradation, clashes of scale where local economic growth leads to the undermining of conditions for life at a larger scale, and how the lack of a control mechanism – like a thermostat – allows the process of irreversible climate change to escalate.

Climate change must be seen in relation to extractive industries, another area in which there is currently a growing social science interest, including studies of coal, oil and gas, but also minerals such as iron, copper and gold (for example, de Rijke 2013; Golub 2014; Kirsch 2014; Pijpers 2014; Luning and Pijpers forthcoming). These studies are ethnographically situated and based on ethnographic research in particular mining areas, yet they convey general insights about the global significance of the mining industry. There is a global mining boom going on, where mines which had been closed have been re-opened, such as in the Zambian Copperbelt. New concessions are being negotiated on every continent, new ownership constellations scouring the planet for mineral wealth. In his ongoing research on mining and accelerated change in Sierra Leone, 'Overheating' researcher Robert Pijpers (forthcoming) demonstrates the multiscalar nature of mining in today's world. The mining operations are owned by transnational companies which have their head offices outside the country; but in order to get started, they first have to negotiate not only with the Sierra Leonean government, but also with local chiefs and landowners. The ultimate success of iron ore mining in Sierra Leone depends on the situation in the global commodities market, the economic situation in China and the efficiency of competitors in countries like Sweden and Indonesia. Locally, the effects of the mining boom have been increased prosperity but also growing inequality, new opportunities and new health problems.

Ethnographic research about mining gives important insights into global economic integration and global inequalities, the sheer environmental brutality involved in open-pit mining, and the clash of scales resulting when a global company extracts profits from a particular area, leaving polluted streams and a toxic slag heap behind when the supply has been exhausted and the company has moved on, producing, at the height of its powers, wealth and poverty, violence and inequalities, as witnessed in the quintessential mining metropolis of Johannesburg, a city of glittering wealth and desperate precarity, of lush, serene suburbs and gritty zones marked by sudden violence.

At the same time as there is a growing interest in mining among social scientists, there has so far been less systematic attention to the broad significance of energy for human lives. It would be an understatement to say that the relationship of fossil fuels to climate change is well documented and debated, but the relationship of energy to the human condition remains understudied, even if it is crucial for an understanding of contemporary modernity, its historical trajectory and its current predicaments. In anthropology, though, there is an incipient interest in the field. A special issue of *Anthropological Quarterly* edited by Dominic Boyer examined 'energopower' (Boyer 2014), and the journal *Focaal* published a special issue on oil and power in 2008, edited by Stephen Reyna (Reyna and Behrends 2008). These collections were mainly about political power, however, much in the same vein as Mitchell's exploration of the politics of oil. The 'Overheating' approach to energy instead emphasises the role of energy as a driving force in instigating accelerated change in areas which are central for human society, notably economic and environmental processes which affect identity and belonging. Accordingly, I see energy through the lens of crises of reproduction and of clashing scales.

Intriguingly, before the very recent resurgence of interest in energy, the most important contributions to an anthropology of energy, published more than half a century ago, left little trace in the discipline – or outside it, for that matter. In a series of original, but long ignored texts, Leslie White (1943, 1949, 1959) argued that there was a direct correlation between energy use and social evolution: 'Other things being equal, the degree of cultural development varies directly as the amount of energy per capita per year harnessed and put to work' (White 1943: 338). By 'cultural development', he refers to increased population density, division of labour and technological sophistication. Although few would subscribe to White's simple evolutionism today, his main insight remains worth examining in the contemporary situation: access to energy is related to social life and social form in fundamental ways. His equating of high energy consumption with cultural advances nevertheless needs questioning; White once summarised his view in a 'law of cultural development: culture advances as the amount of energy harnessed per capita per year increases, or as the efficiency or economy of the means of controlling energy is increased, or both' (White 1959: 56). Even if we bracket the simplistic social evolutionism implied, it might be pointed out that this view represents a dated framing of the question at a time when there is general agreement that humanity needs to reduce its energy consumption. Besides, comparative studies of life satisfaction indicate that there is no simple correlation between the good life and the energy use, or ecological footprint, left behind by a population (NEF 2014).

Energy comes in many forms. Food contains energy which is transformed and utilised by humans (and other species). The introduction of beasts of burden increases productivity in society since they are stronger than humans. A warm climate reduces the immediate need for external energy sources because the amount of energy directly appropriated from the sun is greater in Ghana than in Germany, and so on. Most of the humanly consumed energy in the contemporary world consists of *stored sunlight*, from firewood and charcoal to oil and gas. (Even wind energy and hydroelectricity ultimately depend on the sun.) Solar power, arguably an emerging source of energy in this century, is interesting in that it converts sunlight to energy usable for humans without being stored in a medium such as carbon. While coal and oil have already been stored for millions of years, the main problem for solar energy consists precisely in storage.

For decades, a major concern about stored sunlight (coal, oil, gas) was anxiety that supplies would soon be depleted. In his 1943 paper, White quoted research which estimated that at the time the world had enough oil for another twelve years; we would, in other words, have run out by 1955. New technologies and discoveries have increased oil production many times since then, but the finite supply of fossil fuels has remained a standard example invoked to debunk the image of unlimited good. However, as noted, there now appears to be no imminent end to oil, gas and coal supply, despite accelerating extraction. Unless, that is, the end result is a planet mostly uninhabitable for humans, which many scientists predict.

Against this background, environmental side-effects, from local pollution to global climate change, are at the forefront of current academic and policy concerns with energy, especially with regard to the materially rich countries. Therefore, and also because of fluctuating prices and other reasons, alternative sources of energy are now being explored and tried out across the world. There are high-tech, ostensibly ecologically sustainable cities such as Masdar City in Abu Dhabi (Günel 2011); alternative communes minimising their ecological footprint in a small-scale, but visionary way; legislation and state policies, such as in Germany, encouraging a shift towards renewable sources. At the same time, a major problem for the world's poor is restricted access to energy mainly in the form of electricity, which limits their opportunities to take part in modernity in several significant ways.

* * *

All other things being equal, low-energy societies are more equitable than high-energy societies. Indeed, small-scale hunter and gatherer societies generally become less egalitarian during periods when energy, mainly in the form of food, is abundant (see Wengrow and Graeber 2015 for a recent statement). The 'resource curse' of energy-rich contemporary societies is well documented and has been described, in the case of oil countries, as a 'triple conjunction ... of stagnating social development and poverty; high conflict, often violent; and a tendency toward authoritarian regimes' (Reyna and Behrends 2008: 5). Increased social inequality is also a typical effect of an energy boom, but contemporary societies where energy is generally scarce are also not necessarily equitable, not least because of the uneven access to energy. Since most of us are by now firmly embedded in, and dependent on, social and economic networks on a huge, transnational scale, distinguishing between small-scale and large-scale societies is far less useful, or even meaningful, than it was a few generations ago. Instead, it may be instructive to distinguish between fields or activities operating on different levels of scale. The reason why this matters is that differences in scale may indicate degrees of inequality, and that large-scale systems or activities require more energy than small-scale ones. Moreover, large-scale operations tend to be less flexible, in Bateson's (1972) terms, than small-scale operations. It is easier for a rowing boat to change course than a cargo ship.

The energy-affluent ...

During my fieldwork in Queensland, one of the people I met alerted me to a recent book by Jeremy Rifkin called *The Third Industrial Revolution* (Rifkin 2011), which argues that the combined outcome of solar power and internet communication will transform capitalism by decentralising the worlds of economy and communication, strengthening the autonomy of local communities, and sharing energy through a grid he calls 'the energy Internet'. In other words, he speaks of a decentralisation of energy and information, which, if brought to fruition, would entail the realisation of the vision formulated by Marvin Harris decades earlier.

The woman who spoke about Rifkin's vision was among the few outspoken environmentalists in the industrial city of Gladstone, and she

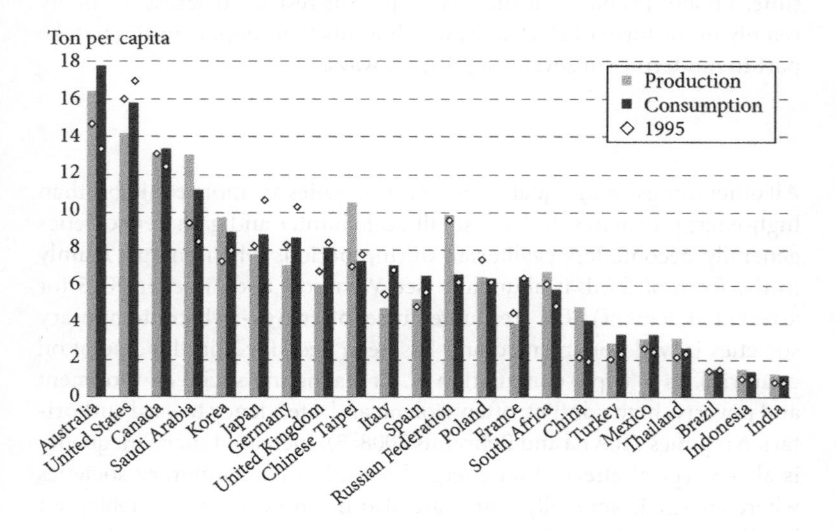

Figure 3.2 Energy production and consumption per capita in selected countries, 2012

Source: OECD (2015).

saw this approach as a way of improving democracy as well as divesting from fossil fuels. 'Obviously, Ergon are never going to encourage this kind of thing,' she added with a shrug, Ergon being the main supplier of electricity in the region.

The model proposed by Rifkin is interesting not least in its combination of small and large scale elements. While anyone can, in principle, be their own electricity producer in a solar-powered regime, surplus power can be distributed to others through a grid on a large scale, indicating how even large-scale operations can be non-hierarchical and decentralised. State power would then be limited to managing the grid, unless it was owned cooperatively.

Soon afterwards, I visited Jan and Karen Arens, two environmentally active residents of Central Queensland. A married couple with adult children, they had moved from Gladstone to a rural property with a bore and a water tank for rainwater, solar panels scattered around the house and a colourful range of useful plants which were struggling somewhat in the dry climate and sandy soil. Unlike most Queenslanders with solar panels, they had gone entirely off grid, meaning that they were completely self-sufficient with electricity. Their energy needs had, in other words, been scaled down to the household level and could be satisfied independently of any other source of energy than the sun. Or so it seemed. On a tour of the property, Jan showed me a very substantial bank of large batteries below the house. They stored sufficient electricity to last four to five days, should the cloud cover be too dense during a rainy season. It did not seem like a lot. In addition, it needs to be pointed out that the solar panels and batteries themselves required some maintenance and occasional replacement. At the same time, Jan says, the implements needed to produce solar power can perfectly well be produced in a carbon-neutral way.

In energy-affluent societies, only people with a personal interest in sustainability and ecology are likely to find ways of increasing their energy flexibility by becoming independent of the fossil fuel industry. Yet, they lose flexibility in other domains, for example through the constraints represented by battery capacity. Others try consciously to live low-energy lives by other means. Alternative communities which produce some of their own food organically and reduce their electricity needs to a minimum (some lighting, charging of gadgets and so on) continue to exist, mainly in the materially wealthy countries, but without making a dent in the hegemonic energy system which remains based mainly on growth in fossil fuel consumption. Nearly 50 years' experience of alternative, green communities shows that global capitalism is capable of absorbing many thousand such green pockets without changing its course so much as a millimetre. There is an obvious clash of scales at work here. The small-scale alternative communities may be self-sufficient in energy, but they can neither be scaled up without losing their character, nor is there any sign that this mode of living will spread epidemically any time soon. This could be attributed to the iron grip that corporate power and centralised, high-scale politics has on people, but it is also an indisputable fact that people in materially affluent societies tend to enjoy their affluence. The small scale of green communities can, in other words, exist in parallel to the large scale of corporate capitalism and mainstream society, without either of them influencing each other to any noticeable degree.

At a higher level of scale, that of national and international politics, a different kind of clash of scales operates. Saving the world can be far easier than saving a community, or so it may seem in politicians' speeches. In the late 1980s, Norway's prime minister Gro Harlem Brundtland became world famous for her introduction of the concept of sustainable development, first presented in the UN report *Our Common Future*, authored by the World Commission for Environment and Development, which she had chaired (UN 1987). She would often speak of the importance of bringing ecological constraints and limitations into economic thinking and planning. However, at the same time, her government was busy expanding oil operations in the North Sea, the then new source of the enormous wealth to which Norwegians soon become accustomed. On the national scale, her main responsibility was towards her voters, and keeping unemployment numbers low and economic growth high were among the main priorities of her Labour government. Besides, as Joseph Tainter (2014) and others have pointed out, the concept 'is not genuinely useful. Fundamentally it borders on tautology: of course sustainable development concerns tending to the future' (Tainter 2014: 221).

A later Labour prime minister in Norway, Jens Stoltenberg, would reveal that this kind of issue is not merely about clashes of scale, but that it also visualises the fundamental double bind in a striking way. Stoltenberg, also of the Labour party, held office from 2004 to 2013, a period in which Norway continued to develop and expand its oil and gas fields, and his party supported exploratory drilling off the Lofoten archipelago in the north, an ecologically rich and fragile area at the conjunction between the warm Gulf Stream waters and the cold Barents Sea. After losing the 2013 elections, Stoltenberg functioned temporarily as a special envoy for climate issues in the UN, and soon began to speak of the urgency of the situation and the need for immediate action. (It was inevitable that frustrated environmentalists should point out that he could have said this years earlier, when he was still running the government.)

At a high level of scale, some energy-affluent societies are involved in compensatory activities in the Global South. Through its oil and gas exports, Norway may indirectly be responsible for as much as 3 per cent of the global CO_2 emissions. At home, the country has a better track record, in spite of the fact that the affluent Norwegians like their beefsteak and are frequent flyers. Most of the energy used in Norwegian households and industry comes from hydroelectric plants. Yet it is commonly known that Norway is a part of the problem, not the solution, when it comes to dealing with the global climate crisis, because of its considerable oil exports. In this context, the centre-left Stoltenberg government sought, mainly in two ways, to balance out some of the detrimental effects of Norwegian oil and gas exports: (i) the directors of the Government Pension Fund, into which most of the state oil profits are invested, are concerned with ethical investments, and have appointed an ethical council which oversees its activities, aiming to ensure that it does not invest in 'unethical' activities; and, more importantly, (ii) the country commits itself to considerable investments in projects aiming to reduce carbon emissions elsewhere, notably in the Global South. The most familiar of these may be the UN-sponsored REDD programme (Reducing Emissions from Deforestation and forest Degradation; see Howell 2015).

The irony is evident: instead of implementing changes at home, notably reducing the rate of oil extraction, Norway pays foreigners to change their behaviour in order to reduce the impact of – among other things – Norwegian oil exports! Stoltenberg, an economist by education, argued that investing in climate-friendly activities in the Global South was far more cost-effective than spending similar sums in the expensive North, notably including Norway. While this is doubtless true on a global scale, the picture looks different at the community level. While Indonesians were instructed as to how to become more sustainable, Norwegians were told that they did not have to change their behaviour.

This duality in Norwegian policy, whereby social welfare and economic growth are closely associated with oil extraction, whereas foreign investments and development assistance aim to reduce carbon footprint and environmental destruction, reveals the central double bind, where growth in energy use and ecological sustainability are desired at the same time, though it is rarely possible to achieve them simultaneously.

A common interpretation of the Brundtland report is that it recommended all rich countries to reduce their levels of non-renewable energy use by at least 50 per cent within roughly 50 years, with

1980 as the base year. The rationale behind this recommendation was threefold: first, the global justice argument implied that energy should be seen as a limited natural resource, and hence to allow for an increased standard of living in the poor countries, rich countries had to reduce their level of energy use. Second, the climate change argument entailed that a reduction in energy use in rich countries was crucial in order to achieve necessary reductions in global emissions. Third, the biodiversity argument implied that even if it were possible to combine global justice and saving the climate with a high-energy global society (meaning that rich countries could maintain a high level of energy use, and bring up poor countries to the same level of energy use), the consequences for biodiversity would be devastating.

A reduction of global energy use does not seem likely within the extant economic and political systems at high levels of scale. World energy consumption doubled between 1975 and 2012. In other words, in spite of the best of intentions among scholars, policy makers and the UN, nothing has happened. Although there are people and communities who set an example through 'best practices', there is no indication that a transition to a low-energy or renewable-energy world society is imminent, although the politicians who signed the Paris Agreement in December 2015 seemed to claim the opposite. This gap between intentions and practices must be understood by reference not only to the double bind, where growth in practice continues to get the upper hand, but also by looking at scale in its different dimensions. Under what circumstances, we may ask, do people act upon knowledge about possible consequences of their actions which are far away both in space and time? It has sometimes been suggested that 'our grandchildren should be granted human rights' (Gaarder 2007), but the operationalisation and implementation of such a principle seems highly impractical in so far as the long-term effects of present actions are impossible to predict accurately. Moreover, as the American sociologist Kari Norgaard (2011) shows in a study from 'Bygdaby', western Norway, in spite of a high level of awareness about causes and effects of climate change, the inhabitants of the town did not change their actions accordingly. They continued to prioritise that which was close at hand, socially, spatially and temporally. They might worry about climate change – Norgaard's fieldwork took place during an unusually mild winter – but preferred to focus that engagement at a high level of scale which would not affect their lifestyles. They might complain that the UN, or the government, or even the oil industry was not doing enough, while at the same time enjoying flying to Southern Europe for vacation, eating imported Argentine steak on Saturday evening and driving their children to after-school activities.

The immediate social and temporal horizon was somehow not seen as directly relevant to the large-scale and long-term challenges facing humanity. There are arguably credible psychological accounts of this tendency, but it also needs to be accounted for culturally and phenomenologically.

First, we should keep in mind that, for 200 years, progress and development were associated with growth in energy use. With the contemporary knowledge of the connection between carbon emissions and climate change in mind, a continuation of this development has now become difficult to defend. Yet, it is not obvious that people committed to a modern view of progress and development should accept what may seem tantamount to reversing the arrow of time.

Second, the feedback loops that would have adjusted current – ecologically unsustainable – action at the local and short-term scale are out of sync with the systemic needs for change: the system demands change now, based on predicted outcomes of 'business as usual' towards the end of this century, while human action typically relates to predicted feedback within a timespan ranging between a few seconds and a few years. The temporal scales differ. At the same time, however, many environmental activists and organisations that work in local communities favour divestment from fossil fuels, and are often frustrated by the clashing scales they experience. In Gladstone, you sometimes hear people complaining that Greenpeace and other large-scale environmental NGOs do not take the local community seriously. 'They are world champions at saving the planet, and pretty good at saving the Great Barrier Reef, but they don't care much about us who are living in this so-called industrial hell,' one said, adding that the green activists in Sydney and Melbourne would probably prefer to close the entire city down. In other words, it is not just policy makers and corporations that operate at a level of scale perceived as alienating and disenfranchising by locals, but ecological activists as well. Sliding up and down the scales is easy to do in practice – we all do it every day – but the implications are difficult to understand, especially so when the different levels of scales have to be reconciled. Sometimes, it is necessary to slide up in order to see the global consequences of your actions; at other times, it is necessary to slide down in order to understand the local implications of large-scale processes.

... and the energy-deprived

It goes without saying that the discrepancies in energy use between rich and poor countries and communities are enormous. Although gaps are

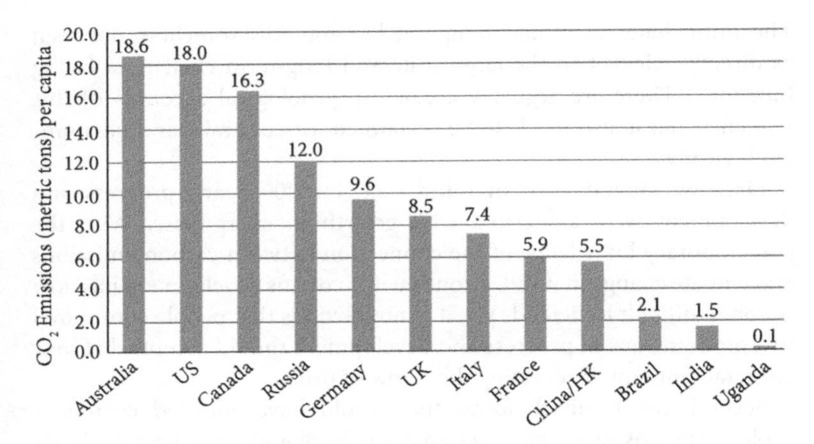

Figure 3.3 CO_2 emissions per capita in selected countries (2014)

Source: World Bank (2014).

closing in some respects – the Germans have become less voracious consumers of energy, while the Chinese have increased their per capita consumption of energy many times over since the late twentieth century – the gap remains huge. In large parts of Africa, electricity is not available or only partly and patchily so. Africans make do without it, just as everybody did before the late nineteenth century; but living without electric lights can be a great inconvenience in an otherwise electrified world. In a comparison between India and South Africa, Akhil Gupta (2015) shows how the poor gain access to electricity in diametrically opposed ways. In the Indian case, people in informal settlements are unconnected to the grid. However, the power companies 'recognize that people need electricity to live in an urban environment. Thus, they unofficially allow the residents of slums to tap into power lines' (Gupta 2015: 561), but not for free. The slum dwellers pay rent to local authorities for their unauthorised access to electricity and, via a complex maze of informal arrangements, the power companies get their money. In the South African case, it has been well known for years that poor urban dwellers who are connected to the grid, but cannot afford to pay the bill, manipulate ('doctor') their meters to gain illegal access to electricity. In a bid to stop this practice, authorities have introduced pay-as-you-go meters; as a result, even people who are connected to the grid cannot use electricity unless they can pay. This, incidentally, was also the situation in the Mauritian fishing village where I did fieldwork in the 1980s. Most of the villagers lived in a housing estate constructed by the government after a devastating cyclone in the 1960s, and they had not only piped water and showers, but also the sockets, wires and fuse boxes they needed to lead electrified lives. However, only a handful of households actually had piped water and electricity, the rest being too poor to pay the rates. As a result, people went to bed early. Night fell at seven in the evening, after which the only sources of light were kerosene lamps and flickering candles. Schoolchildren never did homework after nightfall, and nobody watched TV.

Modernity is never an either–or kind of phenomenon; the illumination of the Enlightenment is never complete (especially, it is tempting to add, if you don't have a functioning lightbulb), and we all live in mixed, or hybrid, worlds where science and religion, bureaucracy and informal relationships, individualism and collectivism coexist. Nowhere is the patchiness of modernity more evident than in many parts of the Global South, where modernity has arrived in a stuttering and incomplete way, at least as seen from an energy point of view. In the Kenyan countryside, energy scarcity is such that you often see vendors at the roadside selling half-litre bottles of petrol. In the south-west of the country, some Maasai have mobile phones which they charge with tiny solar chargers or in the nearest town. People get by, on a shoestring, moving in and out of the large-scale worlds of the modern state and instantaneous communication.

The uneven access to energy typical of the Global South does not just operate at low and short-term levels of scale. Owing to inadequacies in the energy infrastructure, which was initially built to accommodate a pre-'Overheating' population, including a non-air-conditioned middle class, access to energy is often uneven and unpredictable with frequent blackouts. In South Asia and South Africa, planned power outages are known as loadshedding. In a study of energy in the Nepali textile industry, 'Overheating' researcher Mikkel Vindegg (2015) focused on the economic and social implications of loadshedding in Lubhoo, a large village south-east of Kathmandu. Under a loadshedding regime, the timing of the power cuts is announced weeks ahead, enabling factory owners to plan the rhythms of their operation, but production suffers when it is necessary to send workers home for several hours every day. As Vindegg shows, this was a less acute problem in the past, before the expansion of the economic scale to include competitors from neighbouring countries, who may have had larger factories, but also a steady supply of electricity. Since electricity is not new or unknown to the people of Lubhoo, they 'are acutely aware of the opportunities that electricity facilitate' (Vindegg 2015: 13) and experience a loss of possibilities when it is gone. The abrupt shifts between electrified and non-electrified lives in Lubhoo affect not just productivity and competitiveness in the textile industry, but also the availability of labour, since many migrate in search of better and more stable incomes; it affects the middle classes' ability to preserve food through refrigeration, it interferes with activities associated with a modern way of life such as watching TV or doing homework, and creates a peculiar stop–start, staccato rhythm in everyday life. To the extent that people in Lubhoo rely on electricity, their lives are not overheated, rather involuntarily cooled down at often inconvenient times. Limited access to electricity – Vindegg, seeing energy as the lifeblood of modernity, speaks of 'anaemic modernity' – also effectively restricts linear, cumulative development.

Compared to most OECD countries, Nepal does admirably well on a carbon footprint index. According to figures from the World Bank, CO_2 emissions per capita in Nepal were just 0.2 tonnes in 2014, among the lowest in the world, compared to 9.2 tonnes in Norway, disregarding the impact of the oil and gas exports (World Bank 2014). The situation for the residents, and textile manufacturers, of Lubhoo nevertheless comes across as unsatisfactory, certainly from their own point of view. Their problem contrasts sharply with the dilemmas faced by the rich countries, where the challenge consists in finding ways of using less energy; Nepalis are looking for ways in which they may use more. The most likely solution in their case consists in developing some of the country's huge potential in hydroelectricity, not settling for a low-energy adaptation. In other energy-deprived countries, solar energy may similarly be a solution. Since these societies have not committed themselves to high fossil fuel consumption, they are far more flexible, in terms of future energy choices, than those that have a vast fossil fuel infrastructure and rely on powerful resource companies.

Solastalgic meditations

There is no simple solution. Feeding the 7 billion is difficult enough, and it does not presently look ecologically sustainable in the long term; and in the near future, we shall have to feed 9 or even 12 billion. Even if food waste is dramatically reduced, we reduce meat consumption accordingly, and just distribution is ensured, mechanised food production on a large scale is likely to remain necessary unless a large proportion of the world's population reverts to manual agricultural work, and energy must be available to this end. Regardless of its source, energy is needed. There can be no massive return to village agriculture, even if people should, inexplicably, desire to do so. The increased population density and pressure on scarce resources is also the main reason why the 'Paleolithic diet' popular in parts of the middle class is unsustainable. Based on fruit, nuts, meat, honey and so on, this diet might have been viable had the world population not exceeded a few hundred million. Perhaps a middle ground would be feasible. Humanity cannot survive on nuts and meat, and the growing world population will need regular, predictable access to energy in the form of food. This is not the place to delve into the debate about veganism, farmed fish, edible insects and organic food, although it is relevant from an energy perspective; it may well be that the carnivorous era is coming to an end along with the fossil fuel era but, at the present time, it seems distant. The global cattle population grew relatively modestly from 1.3 to 1.5 billion between 1990 and 2012, while the production of chicken meat has increased from less than 60 billion tonnes to 95 billion tonnes between 2000 and 2014 alone. The transition into the global middle class, which takes place on a large scale in fast-changing countries like China and Brazil, is often expressed through the shift to a less herbivorous diet. And who are we, to paraphrase the information officer in the port of Gladstone, to deny them pleasures that we ourselves take for granted?

Food is energy, as is coal, oil, gas and electricity, whatever its source. I began this chapter with the story of coal and modernity, and to this story I now return. Humans need energy, as do other living creatures. In an overheated world, there is not only competition for available energy between and within species; there is also a real sense in which contemporary energy consumption is unsustainable. One is reminded of the slogan from the 1970s environmental movement: we are standing at the edge of a cliff, and are about to take a long step forward. This, in a nutshell, is the essence of the double bind of contemporary civilisation; and in most cases, the main victims of climate change are not its perpetrators.

This is what makes Australia especially interesting. A major exporter of coal and iron ore, but also a country whose inhabitants drive, fly and own air-conditioners, Australia ranks near or at the top of any list of CO_2 emitters. At the same time, the bleaching of corals and the increasing frequency of bushfires, floods and droughts that Australia has witnessed in recent years also make it one of the prime victims of climate change.

These are facts viewed at a planetary scale. However, Australia and Australians also suffer from energy-related overheating at a local scale. It is hardly a coincidence that it was an Australian thinker who introduced, in the early 2000s, the term 'solastalgia', which refers to environmentally induced distress (Albrecht 2005). Glenn Albrecht, an environmental philosopher, explains that '[a]s opposed to nostalgia – the melancholia or homesickness experienced by individuals when separated from a loved home – solastalgia is the distress that is produced by environmental change impacting on people while they are directly connected to their home environment' (Albrecht et al. 2007: S95). As a matter of fact, the concept was developed during a study of local responses to open-pit coal mining in the Upper Hunter Valley in New South Wales. A prime agricultural area, the valley may be reminiscent of the Vale of Glamorgan and similar landscapes in South Wales, which gave the state its name (and where there has also been large-scale coal mining). Coal mining had taken place in the valley for decades, but it began to expand rapidly in the early 2000s, transforming lush, pastoral landscapes into stony dustbowls. The noise from the dumptrucks and excavators could be heard miles away and, as one farmer says, not only did they lose the peacefulness of rural life and the beautiful vistas, but they also lost the starry skies, since the mining pits were floodlit at night. Albrecht and his collaborators have published quite extensively on the subject and, while the concept of solastalgia remains contested (is it really a diagnostic term, or an activist's designation of a form of development he dislikes?), a sense of loss and sorrow associated with dramatic changes in one's natural environment can be found elsewhere as well (Glackin 2011; Jackson 2011). It could be described as a form of homesickness experienced without leaving home, hence operating at a very low level of scale. At the same time, solastalgia fuses the individual level of scale with the planetary: had the world economy not been massively deregulated from the late 1980s onwards; had the Chinese economy not succeeded in satisfying the world's demand for inexpensive goods; had the global thirst for energy not continued to soar, then the growers and livestock raisers of the Upper Hunter Valley would not have felt a sense of loss.

As so often when the scale of the life-world comes into conflict with that of the global economy, people ask themselves who they can blame and what they can do. In the case of the farmers in the Upper Hunter Valley, there has been outrage, protest and demonstrations against the encroaching open-pit mining landscape, but mining continues to expand. In early 2015, it was announced that the Korean company Kepco had purchased nearly all the land in the Bylong Valley in the Upper Hunter, and the pressure to sell is strong. The farmers get a good price, and many are in debt. In February 2015, the state Office of Environment and Heritage (OEH) announced plans for 16 new coal mines in the Upper Hunter, mostly on state land. Both farmers and rural activists reacted to the close collaboration between the OEH and the mining industry, experiencing not only solastalgic reactions but also realising that large-scale projects tend to win in confrontation with small-scale practices. The story of the Hunter Valley is not only an illustration of the double bind of growth and sustainability and of clashing scales; it is also a reminder that, notwithstanding the massive discourse about divesting from fossil fuels and moving towards renewables, coal is still king; coal is still modernity.

* * *

ps: Having finished writing this section, I decided it was time for a break, so I went to the *Guardian* website to read up on the European crises and the latest football news. One of their pages featured an interactive box, courtesy of the newspaper's 'Keep it in the ground' campaign, where you could enter your age to find out how much coal had been extracted and burnt in the world since you were born. In my case, the weight of the coal equalled 43,000 of the Giza pyramids. This is a figure at a level of scale which can only be computed; it cannot be visualised or imagined.

4. Mobility

The uneven rhythm generated by loadshedding in Nepali towns is replicated in many other domains and at different scales. When the French thinker Paul Virilio (2000) said that we had entered an era 'without delays', he was speaking of instantaneous communication, but not of the traffic jams in Dhaka, a city where the number of cars and motorbikes has grown enormously since the late twentieth century, without a comparable upgrading of the road system, revealing a web of subsystems out of sync where private wealth is paralleled by public poverty. Traffic on a multi-lane highway exists in only three main forms, which may serve as a metaphor for some of the implications of overheating processes: free flow, synchronised flow and the traffic jam. Shifting from free to synchronised flow does not necessarily mean that you have to reduce speed, but you have to keep an eye on other vehicles and take care when changing lanes. The transition from synchronised flow to the jam is still incompletely understood, and not only by impatient drivers, but also by scientists building mathematical models of traffic flows. One moment, traffic cruises along at 100 kilometres per hour, and the next moment, it stands completely still, similarly to the loadshedding schedule, but without the intervention of a human being switching off the flow. With the accelerated and intensified contact between formerly discrete parts of the world, not only is communication facilitated, but so are misunderstandings and disagreements. The Danish cartoon crisis of 2005–6, where the publication of twelve cartoons depicting the prophet Muhammad in a Danish newspaper led to riots and demonstrations in many Muslim countries, is a case in point (Eide et al. 2008). Not only was the communication of the crisis overheated, but so were the reactions. It would seem as if the free traffic flow to which Danes had been accustomed – 'Say what you like; only people who are like yourself are going to hear it anyway' – was being challenged by a demand for a synchronised flow: 'Think before you talk, since you don't know who is listening.' Had the diplomatic crisis between some Muslim countries and Denmark escalated, the final outcome would have been a standstill – a gridlock.

Financial crises, moreover, lead to a cooling down of economies: the cranes go silent in front of half-finished skyscrapers, ships are docked indefinitely, the turnover in the housing market grinds to a halt, and the causes of economic crises are sufficiently complex for economists not to be able to predict them.

It should also be kept in mind that accelerated change never implies that everything speeds up. Overheating processes tend to imply the cooling down of places, activities or domains. First, accelerated change never lasts forever. The ex-miners in James Ferguson's *Expectations of Modernity* (1999) reminisce about the time when they still lived in a society committed to modernisation and dreams of progress, when the Copperbelt was humming with activity. Second, there is sometimes a direct relationship between overheating and cooling down: Although the sun shines, some places will be left in the shade, obsolete and forgotten. Third and finally, as Margaret Mead said already in her study of cultural change in Manus Island following the Second World War, 'different parts of a culture change at different speeds' (Mead 2002 [1956]).

It is in the interstices between these processes, which take place at different speed, that cultural lags are produced. These boundaries between rhythms are also the sites of many frictions and conflicts. For conservative Christians and Muslims, everyday life in the twenty-first century may be difficult to reconcile with the rules and regulations embedded in their holy scriptures.

One of the most salient characteristics of the overheated world is increased mobility. Zygmunt Bauman (1999) has said that 'nowadays, we are all on the move', but by this he does not refer to a new form of universal nomadism, which would have been ridiculous; rather that we (or most people) are all affected, directly or indirectly, by accelerated change. In fact, most people still spend most of their lives within a short distance of the place where they were born. As a percentage of the world's population, the proportion of international migrants today is lower than it was on the eve of the First World War (Castles and Davidson 2000), although the absolute numbers are higher, since the global population has more than trebled in the last century. Yet, just as with other tendencies in the contemporary world, different forms of mobility have increased sharply in the last decades. Global air traffic grew by 60 per cent between 2000 and 2012, in spite of temporary dips owing to the 9/11 attack, the SARS scare and the recession in 2007–9. Other forms of mobility, at a more local scale, are also on the rise: according to the Chinese state, more cars were sold in the country in 2012 alone than the sum total of the Chinese car fleet in 1999.

Labour migration, familiar and widespread since the beginning of industrial society, has accelerated significantly in the last decades. The total number of international migrants grew from 65 million to 214 million between 1965 and 2010, and from 1990 to 2010, annual South–

North migration doubled from 40 to 80 million (UN 2014). Although it can be difficult to distinguish between refugees and labour migrants – many migrate for complex reasons – the majority have to make a living in their new country. Millions of international migrants send money to family and associates in their home country, and the rapid growth in remittances witnessed since the early 1990s indicates that more migrants work in the Global North than before, and that they are better paid. In Bangladesh, for example, remittances received rose from less than US$4 billion in 2005 to more than US$11 million in 2011 (Barai 2012). While migrant numbers from South to North grew by 46.9 per cent between 1990 and 2010, money transfers in the form of remittances officially grew by as much as 557.8 per cent; although the methods of measurement are being questioned (Clemens and McKenzie 2014), there is no doubt as to the increasing significance of remittances. In some countries, remittances account for more than a quarter of their GDP, and any sizeable village in contemporary East Africa now possesses not only a top-up kiosk for mobile subscribers, but also a Western Union office.

Notwithstanding the importance of these forms of mobility in the contemporary world, I have chosen to focus on two contrasting kinds of pressure points in this chapter, which discusses mobility (and immobility) from an 'Overheating' perspective. They tell different, but complementary stories about the contemporary world, both growing fast, and both lead to characteristic overheating effects, but the global elites view them in diametrically opposing ways. As the anthropologist Ruben Andersson says in a study of clandestine migration to Europe, 'Globalization … involves … "time-space compression" on an unprecedented scale. Yet while some travelers – whether executives, "expats," or tourists – are celebrated for their powers to shrink distances and connect territories, others are fretted about for the same reason' (Andersson 2014: 4). I shall begin with tourism.

Tourism bubbles

Long before your arrival, you had been reading voraciously about Bali, and you had some ideas about what to expect. Above all, you envisioned a pristine beach and crystal-clear sea ('azure waters', as the tourist brochures say), perhaps lined with coconut palms or an expansive, shady banyan tree in the background. You thought about varied, always delicious and irresistibly aromatic food, hypnotic gamelan music, lush rice paddies, Hindu temples and an ancient, mysterious, multilayered culture offering even a rare glimpse into the pre-Hindu Bali Aga world.

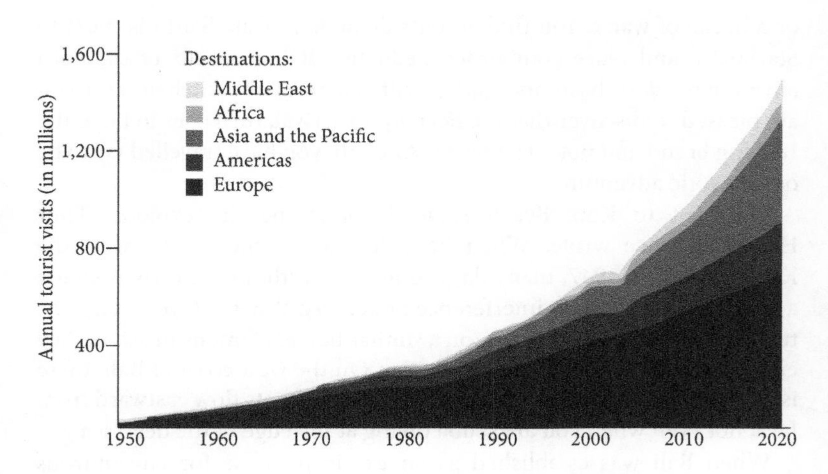

Figure 4.1 Growth in international tourist arrivals worldwide

Source: UNWTO (UN World Tourism Organization, 2014: 14).

The contrast between this imagined Bali and the real existing Bali, as you would soon experience it on the world-famous Kuta Beach, could not have been greater. Neither TripAdvisor, Lonely Planet nor the handful of anthropological analyses of Balinese society that you had perused, prepared you for this encounter. The guidebooks had explained that you needed to negotiate the price of sunbed rentals, and that you should be polite, but firm when refusing the transactions solicited by beach hawkers. However, they had somehow neglected to mention that you were forced to zigzag between piles of rubbish in order to reach the water; and, while wading into the sea, that you would have to pass through a 20-metre-wide ribbon of floating waste, mostly plastic wrappings, before you were ready to start swimming. Incidentally, you might consider keeping your mouth shut during your swim in order not to swallow more water than necessary. Wastewater and sewage from the proliferating tourism facilities nearby tend to enter the sea without taking a detour via a treatment plant. In a place such as this, the poverty of the public sector is just as striking as the private wealth, especially but not exclusively that of the foreign visitors.

After your swim, you may cross the street behind the beach to have a drink and a look at the shops. A whiff of soothing, cool air meets you as you step into the new shopping mall called Beachwalk, where you may be able to save a few quid on international brands if you are interested, but where there isn't a single establishment that sells a local newspaper or a bottle of water. You find an outside table in Cafe Sardinia, next to Starbucks, and place your order. Seductive Italian smells of *calamari olive e pomodori*, basil and garlic drift out from the kitchen, and you are pleased to discover that the beer, upon arrival, turns out to be of the Bintang brand and not a Heineken. After all, you have travelled far to be on an exotic adventure.

The visit to Kuta Beach recalled something the explorer Thor Heyerdahl once wrote. When he sailed across the Pacific with the Kon-Tiki raft in 1947, many days could pass without the crew noticing a single trace of human interference or activity. When, 20 years later, he traversed the Atlantic with Ra, on a similar but less famous mission, they encountered floating rubbish every day. On the west coast of Bali, there is no escape from human waste when ocean currents flow eastward from Java, not even when you are scuba diving at the edge of the deep sea.

When Bali was established as an exotic paradise for adventurous Westerners (including some anthropologists) in the interwar years, the island had about a million inhabitants. It was already pretty densely populated. Today, the population is 4 million and growing. Between the turn of the century and 2014, the Balinese population increased by a million. If you are going somewhere in a car, for example from the west coast to the cultural capital Ubud, you easily get the impression that all 4 million are driving cars or riding motorbikes simultaneously. An hour's drive can easily take two or three hours. The South Sea paradise has become afflicted with constipation.

This story has repeated itself in many places. It is sometimes said that the Balinese culture and way of life has been surprisingly immune from foreign influence, 80 years after Western visitors first began to warn that the island's traditional culture could be threatened by foreign influence. Perhaps Balinese culture has shown admirable resilience throughout the long encounter with Westerners; artists, scholars and musicians since the 1930s, Australian hippies since the 1970s, global holidaymakers since the 1980s. But seen from the perspective of the general trends in international tourism, the story about the lost Balinese paradise is neither unique nor even unusual; and, besides, the gamelan orchestras and theatre traditions of the islands are mainly kept alive by tourists willing to pay for a slice of something old and authentic.

Among the many growth curves pointing steeply upwards, and which have turned the twenty-first-century world into a glowing hot planet, tourism is one of the most striking ones. Granted, world energy consumption has doubled since 1975, and world population growth is not far behind. But in the same period, the number of international tourist arrivals has grown much faster. As summarised by Elizabeth Becker:

> For the first time in history, the U.N. tourism organization celebrated reaching 1 billion international trips in a single year in 2012. The graph line for this travel phenomenon goes straight up: 25 million tourist trips to foreign countries in 1960; 250 million in 1970; 536 million in 1995; 922 million in 2008; 1 billion in 2012. Overall, that represents an annual increase of over 6 percent. (Becker 2013: 17)

As early as the 1930s, well before the publication of Lévi-Strauss's elegy *Tristes Tropiques* – a book which begins with the memorable sentence 'Travel and travellers are two things I loathe' – tourists were worrying about their impact on local cultures. Or rather, particular kinds of tourists were, namely the anti-tourists; that is, those tourists who somehow refuse to own up to their identity and who therefore seek out destinations which attract few other tourists (presumably apart from other anti-tourists). The anti-tourists, recently reinvented as cultural tourists, make up a significant proportion of the total numbers and, in their guise as native Europeans, they have for decades worried about the impact of mass tourism on their own culture. Already in the 1960s, sarcastic jokes were told about American and Japanese tour groups who 'did Europe' in a couple of weeks, who were served standardised 'tourist menus' and rushed through the Sistine Chapel in 10 minutes because they had to catch the bus. But this was just a feeble beginning. Today, the most important pieces of advice, before you travel to Paris or Rome, are about avoiding queues. Buy your entrance tickets on the internet before leaving, they say; or get to the Colosseum before the crack of dawn. Everything else seems to be of secondary importance, but even with pre-paid tickets, crowds are unavoidable. When we made our way through the museums of the Vatican some years ago, the rooms were so thick with people that the risk of pickpocketing was a greater concern than catching a glimpse of the art on the walls. Luckily, Michelangelo's painting is located on the ceiling. Claustrophobics beware.

True to the logic of specialisation and differentiation characteristic of the contemporary world, tourism now exists in an increasing variety of forms, ranging from hiking tours in the Andes to Caribbean cruises, from sumptuous luxury vacations in one of Dubai's enormous hotels to cultural tourism of the post-Apartheid brand in South African townships; from five-star hotels on Hajj in Mecca to scenic boat trips in the Norwegian fjords. Some aim to participate in ecologically responsible tourism, yet they travel by plane to their sustainable destinations; while others prefer a sunbed on a crowded beach, cocktail in hand. Some wish to make themselves useful while travelling to a poor country, and accordingly, there is a thriving industry of orphanages and charitable organisations around the hotels, fake as well as genuine, helping indecently wealthy tourists to feel better about themselves. There are tourists who prefer to visit well-defined places with a legible history and culture, while others opt for non-places, destinations which are as generic as airports. To many, the difference between the original and the copy is irrelevant. Many of those who use the artificial ski slope in Dubai, later assert that they no longer need to travel to the Alps. Others continue to hunt for that which is different, special and authentic. But as the months and years go by, they have to travel ever further away from the nearest international airport or cruise ship port.

Since the turn of the millennium, tourism has changed its character, and has become a prime example of overheated modernity. So many things change and grow faster in this overheated world, which so many see as being *trop plein*. Some may believe that the future is represented in places like Las Vegas and Dubai, desert cities with no other mission than being gigantic consumption reserves, created by mafia money and petrodollars, respectively. A third example might be Cancún, just north of Mexico's *Riviera Maya*, a resort city that was conjured up when the beaches of Florida began to fill up. Americans needed a warm and pleasant holiday destination no more than four or five hours' flight from the cities on the eastern seaboard. They found a perfect strip of sandy beach and jungle on the Yucatán peninsula and set to work. Before 1970, there was not so much as a fisherman's village where Cancún now lies; by 2015, it is a city the size of Leeds, not counting the tourists; and perhaps the most Mexican feature in Cancún is the economic power of the drug cartels and their occasional massacres, admittedly small-scale, in the city.

Laments about the disenchantment of the world owing to increased mobility, the spread of modern institutions and, not least, tourism, are not new, and the expanding niche of anti-tourism, or cultural tourism, draws on exactly this nostalgia. As I mentioned at the outset of this book, already in *Tristes Tropiques*, Lévi-Strauss asked, rhetorically, 'what else can the so-called escapism of travelling do than confront us with the more unfortunate aspects of our history?' (Lévi-Strauss 1961 [1955]: 43).

When the French anthropologist wrote these pessimistic lines, the number of tourist arrivals in the world was about 2 per cent of the present figures. Yet there are important nuances. The truth is that tourism has many futures, and the non-places are just one of them. It is true that cruise tourism has grown explosively since the turn of the millennium, notwithstanding 9/11 and the global financial crisis. In 2000, slightly over 7 million people went on a cruise; by 2014, the number was more than 21 million; in other words, there are three times as many cruise passengers now as there were just 15 years ago.

It is also worth noticing that the building reckoned as the world's tallest at the time of writing, the 829.8-metre tall Burj Khalifa skyscraper in Dubai, was inaugurated only in 2010. It was built by underpaid and overworked South Asians, and the suites and luxury flats are owned – if not actually lived in – by wealthy globetrotters from around the world. The Armani hotel, which fills twelve of the lower floors, is staffed by people from many countries, few of them from the Emirates.

The development of the fishing village of Dubai into a global financial centre and playground has come at a price, and not only in terms of human exploitation and suffering:

> The dead fish can be found floating in the Dubai Creek, victims of the pollution in the [saltwater] river that flows through the city and celebrated as one of its tourist attractions. In 2009 the number of dead fish exploded to more than 100 tons. Most were pulled out and buried in overflowing landfills. (Becker 2013: 196)

There are innumerable niches in the contemporary tourist industry, most of them far less conspicuous than the Dubai hotels; and virtually every one of them is growing. There is, for example, a growing interest among cultural tourists in entering Bhutan, whose authorities have consciously restricted tourist capacity in order to avoid being overwhelmed; there are more cruises to the Antarctic than ever before; African national parks now have far more tourists than elephants, and Venice finds itself in a state of chronic crisis owing to the millions filling up the already cramped city every year in quest of an authentic and unique sliver of European cultural history.

Demographically, the greatest change since the turn of the millennium has consisted – as in many other domains – in the sudden predominance of the Chinese. Since currency restrictions were eased in 2011, millions of Chinese have been able to travel abroad and spend freely for the first time, at the same time as the affluent classes of the country are growing fast. In 2011, the Chinese spent US$72 billion on holidays abroad; in 2012, the figure rose to US$102 billion and, overnight, the Chinese had become the largest spenders, by nationality, in international tourism. Whether this development will lead to a gradual shift towards casinos at the expense of sunbeds it is too early to tell, but the innumerable colleges and universities that give courses in tourism will soon be obliged to add a module of basic Mandarin, if they have not already done so.

The Chinese do not only increasingly travel abroad; there is also a rapidly growing international tourism industry in China itself. As the seasoned travel writer Elizabeth Becker points out:

> by 2020, when China is expected to become the number-one destination, tourism will provide over 10 per cent of China's GDP. To keep up with all those tourists, China is expected to need 5,000 additional new passenger airplanes at a cost of $600 billion. (Becker 2013: 312)

The most dramatic qualitative change in international tourism has been neither growth nor differentiation, but the fact that tourism has in many areas passed a tipping point, where local communities increasingly exist for the benefit of tourists. In the past, it was the other way around. Except for artificial destinations like Dubai and Cancún, tourists tended to arrive as guests in pre-existing local communities.

In the oldest and most visited destinations, the change has been evident for some time. The mass produced *steak frites* served with a glass of *vin de la maison* which is common fare in the pavement restaurants lining the Champs-Élysses is scarcely representative of the diversity of French cuisine, but it has long been established as part and parcel of Parisian culture. In Dubrovnik, which not only fills up with visitors in the summer months, but which is also protected from architectural change by the UNESCO (UN Educational and Scientific and Cultural Organization), residents speak about the feeling of being unpaid extras in an outdoor museum that never closes. Venice has long battled against the pollution and material damage wrought by the urinating bodies and stamping feet entering the city every day. On the Costa Blanca, I once took a room in a hotel where 15 TV channels were available, not one of them Spanish. All of this is familiar, but the tendency has become perceptibly more widespread than before, since the number of tourists has grown 400 per cent in just a generation. No wonder that a major new trend in research on ethnicity concerns the commercialisation of identity (Comaroff and Comaroff 2009) and attempts to achieve copyright in cultural products, material as well as immaterial, in order to be able to commercialise them locally rather than being overrun by large-scale companies, seen locally as cultural pirates (Kasten 2004).

During fieldwork in Australia, I made the acquaintance of several people who had, upon retirement, moved from industrial Gladstone to the Sunshine Coast, a string of suburban settlements in picturesque surroundings some 300 kilometres to the south. One couple, in particular, spoke enthusiastically about the move from a city where there was hardly anything for tourists, to an area where there was scarcely anything but tourism. The Sunshine Coast and Gold Coast, which are located north and south of Brisbane, respectively, have altogether about a million residents, and receive many times as many tourists annually, most of them Australians on holiday. The inhabitants of these places may also sometimes define themselves as tourists. I asked the couple: 'But does it really matter to you that there are so many tourist facilities on the Sunshine Coast – you're not tourists yourself?' The husband answered: 'No, but down there, you can have a lifestyle as if you were a tourist all year.'

When you ask Caribbean or African children what they want to be when they grow up, they sometimes answer: 'Tourist'. In a perverse fantasy story, this might be where the world, or part of it, is heading: The middle classes of the world become tourists, while the rest look after their needs or are employed to give the impression that the visitors have come either to a comfortable non-place or to a real place with a proud history and unique character. Indeed, the English author J.G. Ballard conjured up such a fantasy already in his 1982 story 'Having a wonderful time', albeit with a characteristically dystopian tweak. In the story, British tourists in the Canary Islands slowly come to realise that their delayed flight will never show up, that they have been forever exiled and condemned to live the rest of their lives as tourists.

Obviously, this is a distortion and an exaggeration, but less so today than when the story was written. Many of the North Europeans who move into flats and condos in Southern Europe are superfluous in their home countries. True, they move south by choice, but many depend on social security payments, often pensions, from home (a kind of state remittance) and are, structurally speaking, equivalent to the eternal tourists conjured up by Ballard.

The increase in tourism and tourist-like modes of existence is a paradoxical result of a globally integrated economy, heightened mobility for the well off (but not for the less affluent), striking inequalities and a middle-class culture focused on consumption rather than production. Of course, there are large parts of the world where you are unlikely to encounter more than a handful of tourists at any time. But in the continuously expanding core areas of tourism, from the beaches of Thailand to the shopping streets of central London; from Hawaii to the Kruger Park, the tipping point has been reached. These places are now defined by tourism, not the other way around. The locals have to play their part well in order not to lose their comparative advantages relative to their competitors, and above all, they will soon need to acquire a smattering of Mandarin.

* * *

Global tourism growth, made possible by the abundance of fossil fuels, carries all the main characteristics of overheating. Its growth has been spectacular and continues at an enormous speed, often in a runaway fashion with severe unintended consequences, the most obvious being overpopulation and loss of local character. In Mauritius, the authorities realised, in the 1990s, that the island's attraction lay in the serene, quiet, almost pristine character of its coastline, and therefore passed a law banning buildings higher than the tallest coconut tree. Not all have been equally prescient. As with humanity's exploitation of fossil fuels, some tourist destinations have somehow been too successful for their own good, shifting almost unnoticeably from delightful and exotic destination to crowded, noisy and smelly inferno.

Global tourism is simultaneously an important part of, and made possible by, the deregulation of markets through the globalisation of neoliberalism. Investments by transnational corporations in tourism infrastructure, from hotels to package trips, have been greatly facilitated since the 1980s, as has the mobility of the world's middle classes. There are cheaper flights available, less bureaucratic red-tape, more profits to be gained.

Tourism is also a major factor in bringing the Anthropocene to fruition, through its networks of connections, infrastructural developments and waste production on a massive scale. The often lamented cultural homogenisation allegedly produced by globalisation is evident in tourist centres, which may be seen as 'switchboards', to use Ulf Hannerz' (1990) term, creating connections and comparability, and affecting tourists and their service-providers alike. A tourist hotel follows largely the same global cultural grammar whether it is located in Bali or in Dominica. At the same time, this very homogenisation, or 'flattening', holds out a positive promise of a possible global conversation about inequalities, transnational connections and the diversity/homogeneity axis. Like the global discourse about climate change, or about the rise of politicised identity politics, tourism is a common language shared by a growing part of the world's population. It is not uniform, and different people will have diverging interests and perspectives – to take the most obvious disparity, a housemaid sees tourism differently from the hotel guest – but since it is so easily recognisable everywhere in the world, tourism creates shared templates for talking about humanity not mainly at the level of the community, but on a global scale. Like money, English and anthropological theory, it produces comparability. These incipient global conversations, however imperfect, asymmetrical and incomplete, may be the most promising overheating effect seen from the perspective of human survival.

Tourism operates at many scales and contributes to peculiar forms of scaling locally. The children who state that they would like to be tourists when they grow up have developed a cognitive scale enabling transnational comparisons – restricted in scope, skewed, but still able to generate an imaginary which connects them to metropoles and territories in abstract faraway locations.

Discrete social and spatial scales meet in every tourist destination, sometimes in a tidy and complementary way, but they also clash. On the one hand, there is complementarity between the Maasai women who sell handicrafts outside the gates of a game lodge in Kenya and the foreign owners of the lodge, just as there is complementarity between the fishermen in Playa de Carmen and the hotel chains that buy their catch. The situation is different when big money and small money enter into the same market, as when Starbucks opens a new cafe on a beachfront where locally owned coffee shops are already present. The tendency here, as in other contexts, is that all other things being equal, big money easily wins over small money.

The double bind of tourism, although it is by default part and parcel of the fundamental double bind of growth and sustainability since it relies on high energy use and extensive air travel, also has another aspect, which has been referred to earlier. When a tourist destination has passed the tipping point beyond which it ceases to be a locality to which tourists come, but rather has flipped into a tourist destination with locals providing services for visitors, it is no longer autonomous and self-defining, but is instead assigned a role in a transnational play whose script has been written elsewhere, beyond the direct influence of locals. In practice, this entails compromising local autonomy, or some of it, for the sake of participating in a system of global scale, while the shift from local to transnational scale also reduces flexibility. As many have experienced during large-scale crisis, be it the post-9/11 fear of flying or the 2007–8 global financial crisis, large-scale events influence small-scale conditions for reproduction in ways which are impossible to control or even influence. In the summer of 2015, in the midst of the Greek crisis, local cafes in Athens saw their turnover dwindle as tourists stayed away, due to instability on a higher level of scale. Relying on economic stability at the level of the state, and basing their material reproduction on foreign visitors, the small businesses in question could not simply revert to a pre- or non-tourism adaptation. Integration into higher systemic levels, seen from the individual actor's perspective, may enable enhanced opportunities for profits, but increases vulnerability and reduces flexibility. This principle does not just hold true for tourism, of course; the double-edged sword of integration into higher levels of scale is endemic to global overheating, where localities increasingly depend on the higher systemic levels for their survival.

The global tourism explosion has increased mobility as one of its main conditions. It fluctuates with economic cycles and other factors enhancing or limiting mobility in the global middle classes. A different form of mobility/immobility, which reveals overheating effects at the opposite side of the inequality spectrum, concerns the waves of refugees trying to find peace and security. What they have in common is mobility, but, as Bauman (1998) pointed out years ago, stating the obvious but usefully reminding his readers of the asymmetries of mobility, the tourist is a voluntary traveller who is welcomed when s/he arrives, while the vagabond or refugee is forced to travel and is rarely received with great enthusiasm upon arrival at their destination.

Refugee crises

Some years ago, I spent a week on holiday with my family in Tenerife. This was before the latest financial crisis, and the economy in the Canary Islands was booming. There seemed to be construction going on in every vacant lot, and the Autonomous Community of Las Canarias was the fastest growing region in Spain. Only the Las Palmas football team, languishing at the bottom of La Liga, failed to meet the universal expectations of success and growth generated by the economic boom. On leisurely strolls in the balmy Atlantic air, we saw shopping centres, condos and terraced flats being built; we walked into locally owned groceries selling Scandinavian newspapers and fragrant Dutch tobacco, and one day I read a story in the local press about the African boat refugees who had just landed on Gran Canaria.

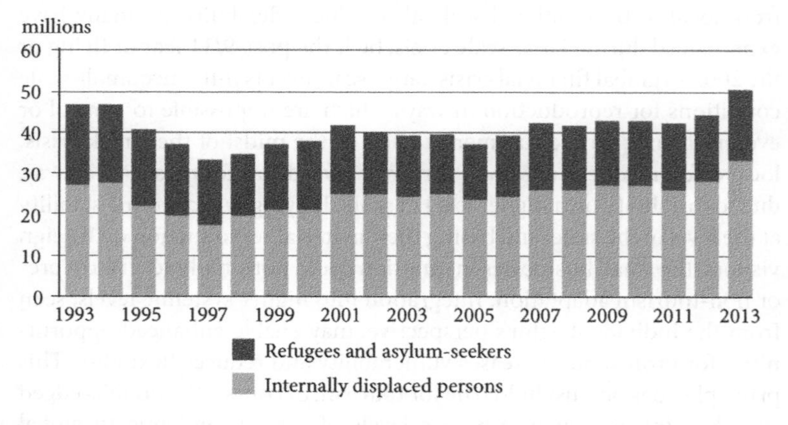

Figure 4.2 Refugees worldwide (millions)

Source: UNHCR (UN High Commissioner for Refugees, 2013).

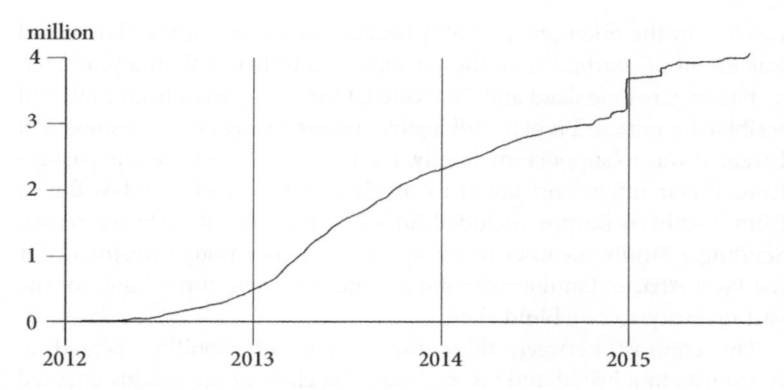

Figure 4.3 Registered Syrian refugees 2012–15

Source: UNHCR (2016).

A striking image of the state of the world emerged from the clash of worlds taking place in the archipelago, and the natives were mainly bystanders on this occasion. Three waves of international migration washed over the Canary Islands: first of all, there were the wealthy North Europeans, sometimes spoken of as migratory birds or climate refugees (within huge scare quotes), who owned comfortable flats in which they spent part of, most of or even all year. Theirs was a world of material affluence, consumer choice and a hint of boredom. Escaping from the drabness of the North European winter, their main project in Tenerife consisted in conjuring up meaningful activities with which to fill their day. Perhaps they occasionally did a spot of golf. The second category was those who actually built the terraced flats and supermarkets, who ensured that the toilets were clean and the drinks chilled. These would-be labour migrants mostly from Catholic countries such as Peru and Ecuador, Romania and Poland, hoping for a more prosperous and secure future for themselves and their children. At the time, their prospects were bright; this would change with the financial crisis a few years later, and many would then return to their countries of origin.

The third category were the boat refugees arriving from West Africa in dilapidated, dangerous, barely seaworthy vessels, and lodged in provisional camps safely out of sight of the tourists on whom the Canarian economy depended. Having completed the dangerous, exhausting and expensive passage, the vast majority would soon discover that they would be returned unceremoniously to the West African port where they had boarded the vessel.

Not everybody made it to the sandy shores of the Canaries. Think, just for a second, about the boat which left Dakar in March 2005, and was lost in the open sea, probably because of engine failure. The vessel was found off Barbados, in the Caribbean, a little less than a year later, its passengers long dead and desiccated. One young man from Mali had scribbled a note in French, still legible, before losing consciousness: 'All I wanted was to support my family. I am sorry.' At the time, the passage from Dakar might cost about as much as a two-week holiday, flights from Northern Europe included, in a comfortable all-inclusive resort. Sending a family member to Europe was an enormous investment for the West African families who did so; and this route turned out, for the vast majority, to be a blind alley.

The contrast between these three forms of mobility seemed to epitomise, in a brutal and visceral way, the clash of life-worlds entailed by a more integrated world. Incidentally, it makes perfect sense that these worlds should meet precisely in Tenerife. In the village of Garachico, facing the Atlantic, a monument was unveiled in 1990. Entitled *El Monumento al Emigrante*, it depicts a man with several suitcases facing the sea. Where his heart should have been, there is a large, round hole. For several hundred years, the Canary Islands were a stepping stone for migrants to the New World, most recently under the Fascist dictatorship which ended only with General Franco's death in 1975. The monument tells the viewer in no uncertain terms that you sometimes have no choice other than leaving your home, but you do not do so without remorse. *El Emigrante* of the monument, at least, has left his heart behind.

* * *

A few years later, it is not boat refugees to the Canary Islands who make the headlines, but the fast growing number of people crossing the Mediterranean, often from a Libyan or Turkish port. Boat refugees boarding in North Africa have perished in the Mediterranean for years, but the numbers are growing. So are the numbers of Syrians, but also Eritreans, Afghanis and others who make it across, but who are facing an uncertain future and, in many cases, will be kept in provisional refugee camps for a long time, waiting for their asylum application to be processed.

The Mediterranean has for many years represented one of the starkest contrasts of prosperity and life opportunities in the world. With the current border regime in Europe and increased mobility in Asia and Africa, the Mediterranean has not just become a frontier between Greece and Lebanon, between Morocco and Spain, but it effectively forms the border between Germany and Congo, between the Netherlands and Sierra Leone. Schengen has lifted the logic of bordering up to a higher scale, from the nation-state level to the EU level. Unlike other borders in the post-Cold War world, notably within Europe itself, the Mediterranean, a network of criss-crossing marine highways since Antiquity, has been reinforced as a border between Europe and North Africa and Asia, rather than weakened. The removal of boundaries is never absolute, and certainly not wherever there are profound and easily noticeable inequalities. In the last few decades, walls have been torn down, the most famous one being that separating the two Berlins; but new walls are being built with great determination and efficiency, along the Rio Grande, through the West Bank, around wealthy gated communities in deeply unequal societies, around the refugee camps in Calais and along the European borders on African soil in Ceuta and Melilla. In spite of the best efforts of Frontex, the European agency responsible for border control in the Mediterranean, it has so far not proved possible to turn the frontier zone of the Mediterranean into a fixed boundary; there is no fence or wall preventing hundreds of thousands of fleeing Africans and Asians from trying, often successfully, to land on a Greek or Italian island.

Those who drown en route are not even counted by European authorities. Europe has been forced to deal with the fact that the Mare Nostrum of Roman times is now becoming a Mare Moriens, a sea of death.

When the internal borders in most of the EU evaporated in the mid 1990s owing to the Schengen Convention, control of the external boundary had to be strengthened. The systematic search for irregular migrants began, and it has been coordinated by Frontex since 2004. In the early years, the main focus was on the route from West African ports such as Dakar and Nouakchott to the Canary Islands; more recently, the itineraries across the Mediterranean have been in focus.*

Although the border was in theory closed, at the time of the mass Syrian exodus, it was more open in fact than it had been for a long time. The collapse of the Libyan state allows easy mobility out of its ports, for those who make it there without being robbed or murdered, and who are still capable of paying for their passage. Talking about this route of travelling as human trafficking, as is common in Europe, does no favours to those who have taken great risks and spent their money to make the trip in a bid to find peace and support themselves and their families in wealthy, well-organised Europe. Unpleasant, overcrowded, dangerous and uncertain as it may be, the journey across the Mediterranean is judged as a risk worth taking for those who do. This says something not only about what they are leaving behind, but also about their hopes for the future.

---

* The following paragraphs build and elaborate on Knudsen (2015).

The growth in the number of boat refugees in the Mediterranean can partly be attributed to the civil war in Syria. The number of people who had fled grew from a few thousand as late as January 2012 to 4.15 million in December 2015. Yet, regardless of future developments in Syria, the waves of would-be migrants are not likely to abate. Many of the displaced people flee from war and oppression, but many are also trying to escape from hopeless poverty, and the means of mobility are now more readily available, with an improved communication infrastructure, more professional middlemen than in the past and an ever increasing awareness – accurate or misleading, as the case might be – of economic opportunities in Europe. In an important sense, the boat refugee crisis in the Mediterranean represents an image of the globalised world of the early twenty-first century; not the most attractive one, to be sure, but one that reminds us of the broken promises of unilinear social evolutionism, the belief that history had a direction and that the rest of the world would sooner or later 'catch up with' the West. The clash of worlds is not heard as a distant thunderclap; it is on our doorstep. That may be the most unsettling aspect of the current refugee crisis for many Europeans. In a not so distant past, Palestinian refugees were mainly confined to their Middle Eastern camps, Vietnamese boat refugees were found in the South China Sea, and the Mediterranean was a playground for North European tourists. That is no longer the case. Refugees from the Middle East became boat refugees in the Mediterranean. Holidaying on serene, picturesque Greek islands, tourists on their way from the breakfast buffet to the beach now had to ride their rented bikes past waiting groups of Africans. In the mid 2010s, thousands arrived on Greek islands every week. Along the beaches of the Mediterranean, you no longer just noticed the rubbish and plastic bags floating to the shore, but you might come across the dead bodies of children and adults. In spite of the best efforts of Frontex and the national coastguards to keep the dividing line crisp and clear, the multitudes of hungry, dirty, exhausted, fearful and humiliated refugees were no longer a faceless mass of people depicted on blurred black and white photos; the new boat refugees meet us up front in overcrowded boats plying the waves of the beautiful Mediterranean on their way to our favourite holiday destination; 'they confront us in high definition, on TV, smartphone, tablet and in real life' (Knudsen 2015, my translation).

Taking on a slightly longer temporal scale, and a larger spatial scale, than is afforded in the debates over the current refugee crises, it is easy to see why the Mediterranean has become a pressure area, and why this is likely to continue given the huge disparities in political stability, wealth and life opportunities between Europe and its neighbours. The demographer Paul Demeny has looked at the development in population between the 25 (now 28) member countries of the EU and their neighbours, from Morocco to Pakistan, and the results are – as Thomas Homer-Dixon says in a discussion of Demeny's findings – nothing less than astonishing:

> In 1950, Europe's neighbors had less than half Europe's population (163 to 350 million); by 2000, their population had almost quadrupled to surpass Europe's (587 to 451 million); and by 2050, according to UN projections, their population will be more than three times larger than Europe's (1.3 billion to 401 million). (Homer-Dixon 2006: loc. 730)

Europe's population is ageing, while almost half the population of its neighbours is under 30. Regardless of acute crises such as the Syrian one, the pressure on Europe's boundaries is unlikely to decrease.

* * *

The UNHCR (UN High Commissioner for Refugees) was established in 1951 to help the roughly 1 million people who had been forcibly displaced after the Second World War. The number of displaced persons in the world was estimated to be about 7 million in 1964, 37 million in 2005, 51 million in 2013 and 60 million in 2014 (UNHCR 2015). At the same time, the number of those able to return to their home countries in 2014 was the lowest in 31 years.

There are various local explanations for the steep growth, but the connection between war and flight is unequivocal: the more violence, the more refugees and internally displaced. Civil war, as in Syria, wars of intervention, as in Iraq, and collapsing states, as in Yemen and Libya, have created new refugee catastrophes within easy reach of the Mare Nostrum. Moreover, these situations have not arisen independently of broader processes and political interventions. The Libyan state collapsed as a result of Western military interventions, and Russia, NATO and the US have actively supported warring sides in Syria. And the boundaries themselves are being reshuffled, destabilised and questioned. The terrorist attack in Paris in November 2015, leaving 130 dead at the hands of militant Islamists, cannot be viewed independently of the engagement of France and other Western countries in the Middle East, which has also seen many civilian casualties.

The Syrian uprising and refugee crisis may, moreover, be linked to climate change. A severe drought lasting several years had crippled the agricultural sector of the country before the insurgency began in 2011, prompting rapid urban growth which was met with little aid from the government. This situation led to rapidly deteriorating relations between the Syrian government and increasing numbers of citizens, creating fertile ground for opposition groups and producing power vacuums, enabling Daesh ('IS') to take control of a large part of the country.

Most of the Syrian refugees are in neighbouring countries. The Zaatari camp outside Amman was the third largest city in Jordan in 2015. At the same time, one in five people in Lebanon was a Syrian refugee. The number of Syrians in Turkey was estimated at 1.6 million by late 2015.

The European response to this humanitarian crisis has been hesitant and indecisive, and European countries have generally accepted only a small fraction of the numbers allowed into Jordan and Lebanon. The main exceptions are Germany and Sweden, which took an open and accommodating position towards the asylum seekers from Syria. Common European policy has so far focused on crushing the so-called criminal networks of 'human traffickers' responsible for taking people across the sea, as well as saving lives; as many have pointed out, there has been a curious mix of humanitarianism and military action emanating from Europe into the Mediterranean in connection with the refugee crisis. There were no concerted plans for resettlement. Rather, the hope seemed to be that the tide could be stemmed, by military force and by interning refugees in temporary asylum centres, until they could be returned. However, the flow of humans across the Mediterranean cannot be stopped permanently by such means, given the enormous disparities in life opportunities. This is a dilemma faced by European politicians at a time when one characteristic overheating effect consists in increased mobility. Determined to keep the external border of Europe intact, European authorities are nevertheless perfectly aware that it is not a 6-metre-high fence with barbed wire on top and watchtowers; it is a semi-permeable membrane, a leaky vessel; and the flows across it are virtually a social analogy to the second law of thermodynamics.

* * *

Staying with natural-science analogies for a moment, a favourite metaphor in early writings about globalisation deserves mentioning, namely the butterfly effect. First introduced by the geophysicist Edward Lorenz (1972), butterfly effects are large-scale effects resulting from small-scale causes, which start a process which expands and slides up the scale ending with a massively noticeable result. As when a butterfly flutters its wings in Trinidad, setting the air around itself in motion, which in turn pushes other air molecules in a particular direction, joining up with a faint breeze and tilting it towards the north, and so on and so forth until New York City is struck by a blizzard. Similarly, small events – a family fleeing religious persecution, a young man searching for a way to support his family – add up to massive transformations of local life and communities, not everywhere, but where their paths converge: at border crossings, in port cities, in refugee camps and, in some cases, in European suburbs. The large size of a refugee camp is a side-effect of many small-scale events which in turn result from large-scale processes (war, religious discrimination, foreign interventions ...), and established refugee camps themselves mutate into large-scale forms of organisation with a complex division of labour, thousands of complementary statuses and so on.

As previously mentioned, the proportion of the world's people who lived in a country where they were not born was higher on the eve of the First World War than it is today. There are nevertheless two major differences: at the time, many of the international migrants were white, and they were often welcomed by the political authorities, whether colonial or not, where they arrived. Second, there are now more than three times as many of us than there were in 1914, most national borders are less negotiable (certainly for non-whites), and the issues raised by governments that consider receiving refugees are framed as problems of integration, instead of asking about the positive contributions that the incoming groups might make. Many OECD countries have quotas for highly qualified immigrants from non-European countries, who are channelled quickly through the immigration bureaucracy since there is a perceived need for them. No matter where you migrate, and from where, you will – all other things being equal – be far more welcome if you are capable of producing and/or consuming efficiently than if you merely need protection. In this way, neoliberal ideology has become an integral part of the contemporary migration regimes (Aas and Bosworth 2013). Migrants who are not seen as economically productive, are unwanted, although some are still granted asylum for humanitarian reasons.

Refugees are mobile, but their mobility is uneven and punctuated by long periods of immobility, almost in the manner of Nepali loadshedding: systemic constraints prevent them from moving freely and without impediment. Much of their time 'on the move' is spent waiting.

There is no typical itinerary for a refugee from, say, Syria or Eritrea on their way to Europe. Syrians may travel via Turkey or via Egypt, the former being by far the more expensive option, and not necessarily safer. In both countries, they rely on middlemen to help them cross the border into Europe, whether by land (Turkey) or by sea. Stories abound about prospective refugees who have lost their money because a middleman was untrustworthy. The numbers of those whose plans are thwarted even before they set foot on a boat are unknown; what we do know is that tens of thousands actually make their way across the Mediterranean and have so far ended up in a state of indefinite waiting.

* * *

Paradoxically, the departure resulting from an overheated and untenable situation at home leads to indefinite, long periods of cooling down in the form of inactivity and waiting. Waiting begins already before you leave home. Although you decided to leave long ago, you have to wait. Perhaps today is not a good day, perhaps tomorrow or next week would be better. Perhaps you have to wait for relatives who are meant to accompany you, or relatives who are meant to send gifts or money with you, or who just want to say goodbye and give you their blessings. Waiting begins long before the journey. After departure from your town or village, the slow movement towards a better future is interrupted by long and often indefinite periods of waiting. The vehicle taking you towards the coast has a mechanical problem or a flat tyre. The middleman takes your money and walks out of the dilapidated flat you are sharing with 20 other clandestine migrants, leaving you to wait for hours or days, if he returns at all. If you eventually make it to the port city, you spend hours, or days, or weeks, waiting for the boat. When you see the boat, your heart sinks. It is even more dilapidated and unsafe than you had imagined. But there is no turning back now. You have invested too much in this moment. You then spend days or weeks waiting in the boat during the unpleasant and dangerous crossing, but none of this can be compared, at least quantitatively, to the period you then wait to have your application processed. It can take weeks, months, years even, for the state to decide on your fate, and for the boat refugees crossing the Mediterranean, prospects are far from bright at the moment.

These periods of waiting, punctuating the mobility which gives you the feeling that you're going somewhere, leaving your tormented past behind a little bit further with every step, can be a recipe for despair and frustration. Boat refugees in another part of the world, namely Nauru and Manus, where Australia has placed some of its internment centres, spend much of their day waiting in queues for basic services such as food, water and toilet facilities.

Waiting, in other words, does not only express power asymmetries, but is also important in itself, as a social fact. Anthropologists in the field often sit outside public offices waiting for important documents, or informants make us wait for them and we duly oblige, knowing that we are there at the mercy of our informants, not the other way around. With refugees, their waiting represents a tacit acceptance of the fact that *scarcity of time is a scarce resource*. They have left everything behind and have nothing left, not even the right to a full agenda. Their time is abundant and empty. Perhaps, in the camps and reception centres on the European shores of the Mediterranean, they sleep more than they used to. Life is placed on hold. While waiting, they imagine a future which may be as bright as it is uncertain, a directed and cumulative form of temporality, quite the opposite of their present situation.

This involuntary cooling down at the border, or at the port, or in the detention camp, suggests that refugees are excluded from participating in overheating processes. Theirs is largely a passive time, until they may eventually be able to return and patch up their broken lives, or find a new home, struggling to adjust to climate, language, jobs and food.

Compared to the tourists, not only are refugees poorer and moving against their will; they are also anomalies in an overheated world; as liminal, unproductive persons, they are matter out of place. They contribute little to the runaway processes of economic growth or to the central double bind between fossil fuel addiction and ecological sustainability. Their personal flexibility is, in principle, almost unlimited in terms of their destination and later livelihood. Synnøve Bendixsen (2015) describes the lives of undocumented Palestinian refugees in Oslo whose main wish is to be allowed to do something useful; to work, learn the language, develop relationships. Yet they are only allowed to wait, or rather, await their imminent deportation. So although the global refugee crises are causally connected to other overheating processes – identity politics, economic crises, environmental destruction – the lives of refugees may serve as negative examples; they are being 'warehoused' in detention centres or, in many cases, camps in a neighbouring country, without a real stake in the overheated world. So in spite of their ultimate cause being an overheated situation – fast changes making life at home economically or socially unsustainable – the streams of refugees represent the opposite of overheating. They are surplus matter; superfluous, unnecessary and inert, they are neither efficient consumers nor producers and can therefore be dispensed with.

As pointed out by my 'Overheating' colleague Elisabeth Schober (2015), Marx spoke about surplus labour and relative overpopulation by referring to those who, permanently or temporarily, are not needed in the production process. In this sense, refugees without wage work may be seen as irrelevant and superfluous to capitalism, but also to societies which struggle to integrate them socially and culturally. At the same time, surprisingly many get by after leaving the refugee centre, sometimes by creating their own 'parallel societies' relying on informal work and connections, superfluous in the eyes of the state, but not in their own communities. Like so many others in the contemporary world, they question and challenge boundaries. They disappear into the labyrinthine networks of the swelling cities of the world.

5. Cities

Towards the end of February 2014, the infamous Beijing smog reached its densest level on record, and foreign correspondents wearing facemasks were seen on TV around the world while reporting the crisis. The number of foreign visitors to the Forbidden City and the Great Wall declined dramatically, and the authorities advised people to stay indoors if they could – easy for tourists, more difficult for construction workers or a delivery boy on a bike. 'The smog has some similarities with a nuclear winter,' said a Chinese researcher I saw on Australian TV, explaining that agricultural production would suffer because not enough sunlight reached the plants.

The WHO (World Health Organization) warns against breathing air that contains more than 35 microgrammes of PM 2.5 particles per cubic metre. (Never mind the technicalities.) In Beijing, the levels stayed well over 500 for days. The main cause of the smog, incidentally, was not the fast growing car fleet in the city, but factories and power plants burning coal in and around the growing metropolis.

Around the same time, the Australian press reported a violent attack on an Australian-owned (but conveniently outsourced) asylum centre on Manus Island in New Guinea. Many of the unarmed asylum seekers were seriously injured, and one was killed. But, as an Australian who lives in New Guinea remarked to me, he would scarcely have been safer outside the asylum centre than he was inside it. Papua New Guinea is one of the most violent societies in the world and, especially in the capital Port Moresby, it can be exceptionally dangerous to move about, especially for foreigners, unless you know exactly where you are going and how to get there. The city is not large (about 400,000), but it has grown fast, and many of those 400,000 have a lot of empty time on their hands. In the previous generation, they would have engaged in horticulture in a village, watched over by older relatives. Today, many of them instead become *raskols*, petty gangsters, in a city where the unemployment rate is officially 60 per cent, but in reality probably even higher.

In a world overheated by, among other things, large-scale investments, mining, population growth, mobility, fast growing agrobusiness and infrastructural developments, space is scarce. Places that only a few decades ago were peaceful and slow to the point of seeming sluggish, are

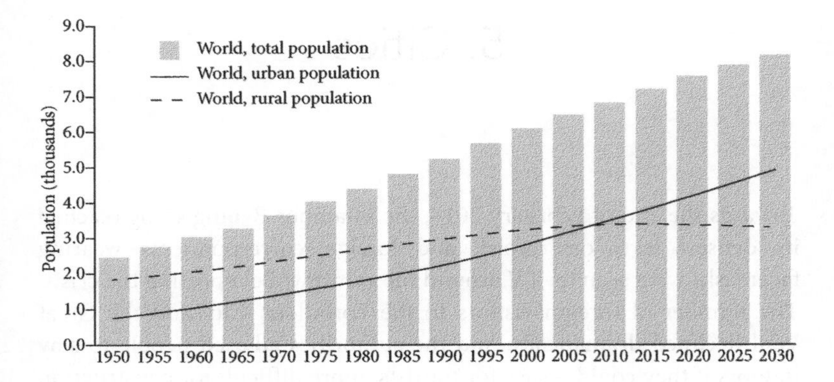

Figure 5.1 Urbanisation in the world since 1950

Source: UN (2014).

now characterised by frictions, tensions and breathless competition for jobs, money, sex, power and other people's attention. If it is true that 'the world is on the move', most of its inhabitants are not moving into refugee detention centres, but into towns and cities. That is beyond dispute.

* * *

São Paulo has been a cosmopolitan crossroads and a dynamic metropolis for more than 100 years, being Brazil's main port for exporting coffee. Today, the city has more than 20 million inhabitants. It is the largest city in the Southern Hemisphere, and a growing number of the residents own a car. On Friday afternoons, the sum of the length of the traffic jams in and out of the city often reach more 200 km. On a Friday evening in November 2011, a total 500 km of traffic jams was measured in the city, minor roads included. It may well be a world record. Paulistas are proud of their metro, but the lines extend only for 75 km. Many car drivers spend the time in the car watching TV, applying make-up, shaving, talking on the phone or sending email, while watching the bumper ahead. A lawyer who spends between two to three hours getting to work every day, says that he feels like a prisoner.

The traffic jam is one of the most poignant and telling images of overheating, starkly revealing a paradoxical side-effect of acceleration: a technology of speed and modernity such as the car helps people accelerate until it reaches a point where it flips into its opposite, namely slowing down. The jam also points to an accelerated growth that few had predicted, and which removes flexibility because it grows without changing the context in which it grows. In Dhaka, the capital of Bangladesh, a city which has grown from a few hundred thousand to 18 million in a generation, only 7 per cent of the urban area consists in roads, while the common figure for a large city is about 25 per cent. (Yet, even in a 25 per cent city like Kuala Lumpur, slow-moving queues are common.) In 1970, the amount of traffic – bikes, rickshaws, scooters, buses, some cars – may have been just right for the infrastructure in place. Not so today. Often one is better off walking, but the roadside is crowded, too. The traffic problems experienced in many fast growing cities, from Jakarta to Mexico City, are also suggestive of the neoliberal hegemony and its limitations. While private prosperity is growing in some segments of the population, enabling many people to buy their car, public poverty prevents local authorities from building new roads. In Bali, the few buses you see, usually stuck in a jam, are dilapidated and overcrowded.

* * *

Although world population, energy use and mobility have increased hugely since the beginning of the Industrial Revolution, urbanisation has proceeded at an even faster pace. Like the other tendencies I discuss in this book, it is as if it has shifted to a higher gear since the early 1990s.

At the time when Napoleon was exiled to St Helena, people were mainly rural, and most of them spent much of their time producing food. In large parts of the world, from Melanesia to Central Africa, towns as we know them did not even exist. Yet the city as such is ancient and central to the cultural history that eventually brought us into the Anthropocene. It is younger than agriculture, but older than the state. Whether we look at Luxor, Babylon, Athens or the Aztec capital Tenochitlan, or for that matter London, Amsterdam or Paris in more recent times, the city has always been a source of increased social complexity and cultural creativity. In Leslie White's basic evolutionary model, where social evolution was measured as a function of the amount of energy harnessed by a population, cities were only possible after an energy threshold enabling a substantial proportion of the people to engage in non-agricultural activities was reached. In the city, the division of labour was advanced. Since townsfolk were exempted from food production, they were traders and scholars, scribes and servants, thieves and prostitutes, politicians and bureaucrats, soldiers and shopkeepers, butchers and bakers, priests and philosophers. Cities were also cultural crossroads, absorbing influences from outside as well as sending their own people into the outside world. They were cultural switchboards and laboratories for social reforms. They could also give refuge to those who needed to get away from the rural life. *Stadtluft macht frei* – city air makes free – was not just an airy slogan in the German medieval cities, but had a material dimension in so far as feudal vassals could escape serfdom if they could prove that they had lived in a town for at least a year and a day. These towns were small, and they grew slowly. Even today, the south German city of Tübingen, founded before Charlemagne, has just 85,000 inhabitants, and a third of them are students.

Perhaps city air still makes free, but to fulfil its promise, it arguably ought to be cleaner than the Beijing air and more opportunities for sustenance should be offered than in Port Moresby. The urban growth taking place in the Global South today is bewildering and unmanageable, violent and risky. Most of the global population growth takes place in urban areas, to a great extent in informal settlements or slums. While European cities grow at the modest rate of 0.13 per cent a year, the figure for Southeast Asia is 3.4 per cent. The cities in the South, including China, currently grow by about 1.2 million persons a week.

Two hundred years ago, less than 10 per cent of the global population was urban. One century ago, the percentage had doubled. In 1960, about a third lived in towns and cities, and in 2008, the UN declared that more than half the world's population was now, for the first time in history, urban. While this could doubtless be seen as a threshold moment for humanity, it should be kept in mind that the definition of urbanity varies. In Denmark, any settlement with more than 200 people is classified as urban. In that southern country (keep in mind that I am Norwegian, and identities are relational), this means having a village with a shop, at least two bus stops and a licensed cafe. In the US, the lower limit is 3500, while the Japanese regard anything below 30,000 as not quite urban. However, as pointed out earlier, scale is not merely about size, but about social organisation. The Indian bureau of statistics understands this and has, accordingly, defined an urban settlement not in terms of numbers but with reference to social complexity: for an Indian settlement to be reckoned as urban, more the 75 per cent of the adult males have to work outside the agricultural sector. Since the informal sector is very considerable in Indian cities, this probably means that far fewer than 25 per cent get their sustenance from farming or fishing.

At the abstract, statistical level of scale, it is easy to be astonished by how strikingly fast the cities in the South grow. Two hundred years ago, very few African people lived in cities. By 2015, over 450 million did (more than the total population of the continent in 1975), but by 2050, the number is projected to be 1.2 billion. The urban proportion is poised to increase from 37 per cent in 2015 to 55 per cent in 2050. In Asia, the number of urbanites will increase from 1.9 billion in 2015 (more than the world's total population in 1915) to 3.3 billion in 2050 (more than the world's population in 1950). In 1970, the number of people who lived in so-called megacities (defined as cities with more than 10 million inhabitants) was 39 million; in 2015, the number was 380 million, and in another ten years, it is likely to exceed 600 million. At the time of the First World War, fewer than 20 cities had more than 1 million inhabitants. At the latest count (2016), the number is estimated to be 450 (UN 2014).

And I could go on. Of all the overheating processes witnessed in the early twenty-first-century world, urbanisation is one of the most striking ones, possibly the contemporary development with the greatest implications for social organisation.

The city still offers ample opportunities for the realisation of individual dreams and projects, and, by dint of its vast occupational differentiation, it represents a hugely increased flexibility at the level of the individual compared to the situation in the country. However, cities increasingly suffer from constipation, like a person who has eaten too much too quickly. Or like a road designed for a thousand cars an hour, increasing the speed and spatial flexibility of their drivers enormously, where there are suddenly 10,000 cars stuck in gridlock. Most of the 6.4 billion people who are expected to live in urban areas in 2050 (compared to the 3.5 billion who did so in 2015) will lack access to many of the resources that made urban life attractive just a generation or two earlier. Gigantic, continuous strips of informal settlements emerge in Brazil, India, China and West Africa without even a nearby urban centre to relate to. They are just slums, not slums appended to a glittering city. The residents of these areas are likely to wait for a long time before the council decides to give them a metro line, libraries, employment offices and public swimming pools.

* * *

Before the Second World War, most African towns were small administrative centres and commercial hubs. By now, the sleepy trading posts of yesteryear have morphed into sprawling cities with more than a million permanent residents; humble crossroads have acquired taxi stands and shaky multi-storey buildings, outdoor markets have become swallowed up by makeshift, evolving informal settlements impossible to map statistically. This growth takes place without an accompanying upgrading of the infrastructure, with few or no municipal amenities, with chronic shortages of basic services such as water and electricity.

Nouakchott, the capital of Mauritania, could be described as a 2 million city with an infrastructure designed to satisfy the needs of 50,000.

Before 1958, when the French established their colonial headquarters in Nouakchott, unaware that the colonial enterprise would end only a few years later, there was already a small fishing village there. Twenty years later, the city had about 150,000 inhabitants. By 2015, its population may have been 2 million. What has lured people to the outskirts of the capital is not the dream of freedom and prosperity, as the case might have been in Tübingen in 1450, but a long and debilitating drought and armed insurgencies in the desert. Many continue to live as nomads in the city, in tents, but with few or no animals and little or no income. Whereas the large and growing middle class of São Paulo spend much of their day waiting for the cars ahead to move on, poor people in Nouakchott spend much of their day in other queues, waiting for water and food.

In many of the cities in the Global South, there nevertheless exists a labour market for the hopeful arriving from impoverished rural areas. This could be the case, for many, in the fast growing metropolis of Addis Ababa, or the epicentre of African hopes and dreams, Johannesburg. The labour market with vacant niches for the typical rural migrant is mainly informal, unofficial and requires a great degree of versatility and flexibility for those who take part. They should probably not worry too much about health legislation either. One of the most infamous examples in Africa is the part of Accra called Agbogbloshie, on the western outskirts of an almost continuous strip of informal settlements stretching from Benin City in Nigeria through Togo and Benin to Ghana; a 600 km long stretch of coastline inhabited by about 50 million people.

Agbogbloshie, locally known as Sodom and Gomorrah, was an uninhabited swamp north of central Accra until the late 1990s. The area was then turned into a rubbish heap for electronic waste from the Global North and became the home of thousands of immigrants from northern Ghana. The obsolete computers, scanners and printers piling up there are often classified as 'development aid' by the senders, since the rich countries in the North have banned dumping of electronic waste in the South; but in practice, they are useless. However, they offer a livelihood for the residents of Agbogbloshie, who sell metal parts, some of them precious, carefully removed from the discarded office equipment. The work is unhealthy and dangerous, but makes it possible for a diligent boy to earn as much as 4–5 euros a day (Stade 2013).

Different, but complementary to the traffic jams of São Paulo, the economic system in Sodom and Gomorrah also provides a sharp image of the explosive urbanisation taking place in the Global South. A typical life in Agbogbloshie is cramped and smelly – the suburb has no sewerage or water piping – and bases its economy on waste from the North. A considerable part of the urban population growth in today's world now takes place in places resembling Agbogbloshie, in informal, temporary settlements where the smell of rot and decay never ceases, and where life is risky and short.

* * *

In the space of only a few decades, the relationship between city and countryside has been turned on its head. In most of our history, there were no towns or cities. Humanity has mainly lived scattered in small communities, sometimes fluctuating between tiny and somewhat larger depending on seasonal food supply, where life unfolded cyclically, year after year, generation after generation. Later, cities and states appeared, but the city was for the few. Today, rather, the rural life is about to become the exception. Population pressure, depleted soils and *de facto* land grabbing by mining companies or agricultural corporations squeezes millions into the nearest city every year. For those who travel to town in order to realise their dreams of prosperity and to enjoy the many good things the city has to offer, the downsides soon become visible. Among the cities that experience the fastest growth in the world are Kabul, Lagos, Chittagong and Bamako. Their new residents are unlikely to be able to rejoice in the urban comforts taken for granted by the global middle classes. Yet, the alternative, for many, might have been worse.

People will continue to get by, even in the uncoordinated, improvisational new megacities. Yet, an increasing number of people will be vulnerable in ways that make the traffic situation in São Paulo appear as a luxury issue for the privileged. The economic niches carved out for survival become more diverse, but also more competitive. Food and water must be transported further and further as the cities expand. Pollution affects both the quality of life and longevity. The authorities often, willingly or for lack of choice, turn their backs on the hordes who settle inside the city walls without warning or invitation. When space is getting scarce, we lose flexibility and, as I have shown with respect to other domains, reduced flexibility in material life is one of the main effects of overheating. There can be no return to village life or to the well-organised market town with its high levels of trust and an established division of labour. The multi-million strong city of the Global South in the twenty-first century, however unsustainable in both social and ecological terms it may be, will prevail in the foreseeable future. It must therefore be humanised and made more flexible, since it cannot be conjured away.

Urban life and the parameters of overheating

Urban growth in the Global South is an illustration as well as a main expression of the Anthropocene. People are coaxed into the cities owing to deteriorating conditions in the country, as well as the temptations of the most sophisticated and complex social arrangements ever created by humanity. There are push factors as well as pull factors involved, but both are connected to current human transformations of the planet, rural and urban. To what extent the rapid urbanisation in the South can be said to be an outcome of neoliberal ideology and practice is a question discussed at length by others (see for example Sassen 2014), and it is not a simple one. In theory, accelerated urbanisation would have been possible under other kinds of ideological regimes as well. The necessary conditions are fast population growth, increased productivity in agriculture and reduced possibilities for sustenance in rural areas, along with promises or expectations of economic opportunities in the city. The escape from rural idiocy, as Marx put it in his typically blunt way, into a more complex world entails expansion of the social as well as the cognitive scale, no matter what the hegemonic ideology. Yet the growth of cities is *de facto* a result of the neoliberal hegemony. The privatisation of land, in many countries a result of compliance with demands from donors and international financial institutions, has led to the expulsion of people whose relationship to their land was founded in custom rather than title deeds. Much of the current urbanisation results from accumulation by dispossession, David Harvey's (2005) term for a family of practices from privatisation of public assets to land grabbing and property speculation.

Dispossession is not a goal in itself; it has to result in accumulation in order to be worthwhile, and three main drivers are mining, agribusiness and property development. The deregulated and thus increasingly transnational economies typical of the world today, along with the fast growing need for minerals and the spread of centralised agribusiness enterprises in the South as well as the North, has made life in the countryside more difficult than it used to be. For many, that is; not for all. Some rural people can make small fortunes by selling concessions to mining companies or industrial food producers, or make a living as their employees; but many others will be left with nothing.

The extraordinary growth of cities is a classic kind of runaway process: Urbanisation, which for centuries worked for the benefit of rural and urban people alike in many parts of the world, has recently grown in uncontrolled, unplanned and unpredictable ways, leading to many kinds of unintended consequences. Commenting on the contemporary megacities, Homer-Dixon points out:

> In fact, compared with ancient Rome, they are enormously more socially and technologically complex. This is not a bad thing in itself, of course, but it does mean that every minute of every day these cities suck from their immediate hinterlands and from regions far beyond almost incalculable quantities of high-quality energy. (Homer-Dixon 2006: loc. 846)

While this is doubtless true, this is not the first time in history that cities have grown fast. In an important historical study of societal collapse, Joseph Tainter (1988) argues that increasing transport costs and administrative complexity were core factors leading to the collapse of the Western Roman Empire and the Aztec empire. As cities grow, the food producers necessary for the citizens' survival become both more numerous and more distant when there are no technological improvements affecting food production and distribution. In the case of the Roman Empire, Tainter's view, shared by some but not all specialists in the field, is that the colonisation of the remote and ultimately unprofitable province of Britannia was one of the most draining and unsustainable expansion projects of the Roman Empire (along with the comparable venture in Dacia, in present-day Romania); not only did soldiers, scribes and traders have to travel to that foggy, inhospitable island off the European coast, but there was also a constant need to import marble, olive oil, wine and Roman architects and builders to an island which lacked the most basic of infrastructure and offered little of economic value in return.

Urban growth may increase flexibility at the level of the individual, at the same time as flexibility at a societal scale is reduced. As a city-dweller, perhaps you may change your source of income, move to another house or shack, shift your children to another school, change your dietary habits and choose between lots of other alternatives which did not exist in the rural society; but modern urban societies as a whole are far less flexible. They have committed themselves to a high level of energy consumption and have developed a complex network of interlocking social networks and activities. A week-long blackout would plunge even a city like Kinshasa into chaos, although many of its inhabitants don't have electricity in their homes. In the rural hinterlands, the less complex social arrangements entail a far greater autonomy in relation to energy supply and sources of food for survival. A chain is no stronger than its weakest point, and the longer the chain, the more likely it is to break.

While systemic flexibility is weak in urban societies that produce neither their own food or energy, it is much improved in the peri-urban regions lining the outskirts of and interstices between many cities. Here, food is still produced, alongside a plethora of services and manufacturing. Their flexibility is therefore higher; these areas tend to be more resilient than the cities as such at a time of crisis, since they have more options and more 'unused alternatives for change', at least in the short run.

The city is effectively an embodiment of the central double bind. All other things being equal, the more it grows, the less sustainable it becomes. Pollution and waste issues, the risk of epidemics, the dependence on complex supply lines and services organised at a higher level of scale do not just reduce the systemic flexibility in the city but also contribute to increasing its ecological footprint. It is not obvious, however, that urban life should be less sustainable than its rural counterpart. By living close together, people need less transport, and services can be concentrated in smaller areas. The main issue concerns the ways in which their energy, often in the form of petrol, electricity and food, is being produced and makes its way to the city. The problem, from the perspective of the double bind of growth and ecology, is not about urbanity as such, but the way it is organised and articulated with energy production and distribution.

Cities are about complexity and diversification; they are large-scale systems *par excellence*. That does not mean, naturally, that the small scale ceases to exist in a large city, but that the social scales are interconnected in ways which did not exist in rural society. Network size is hugely expanded through a move from village to city; many of the connections are single-stranded, but close and intimate relations continue to exist alongside the weak ties, which in turn connect people to other networks. In a word, an average person knows far more people in a city than in a village, which is a main indication of its higher level of social scale. In addition to the person-centred networks and their permutations, which form the basis of a social organisation at a high level of scale, city people rely on large-scale systems for necessary services and material survival; and, as regards cognitive scale, your world expands through exposure to other worlds at the many interfaces connecting people from various origins in cities. Of course, inequalities, which may be moderate or huge depending on the city, insulate groups from each other, as does the tendency to build gated communities protecting the rich from the poor (Caldeira 2001), yet the complex urban life usually corresponds to an accordingly complex cognitive world.

As pointed out above, it is exactly the complexity and high scale (referring to both size and differentiation) of urban civilisation that provides, at the same time, its wealth and vibrancy, and its limitations, sometimes its ultimate downfall (see Tainter 1988 for historical examples). The city can be likened to the kind of complex logistics involved in constructing a piece of sophisticated machinery. Even if 99 out 100 subcontractors deliver on time, the machine cannot be completed until the hundredth piece arrives. This is the main way in which complexity increases vulnerability. The butterfly effects generated by small events (as when a single signal error in the railway network makes 100,000 office workers late for work) bring to mind a children's verse about pre-modern butterfly effects: 'For want of a nail, the shoe was lost/For want of the shoe, the horse was lost/For want of the horse, the rider was lost/For want of the rider, the battle was lost/For want of the battle, the kingdom was lost/And all from the want of a horseshoe nail.'

We now move to considering two specific aspects of two kinds of cities. Just as I juxtaposed and compared refugees with tourists in the previous chapter, I contrast the informal sector in the Global South with the superdiverse city in the Global North here, in a bid to provide a broader picture of overheating than would have been possible with just one kind of case. On the basis of the foregoing discussion, I ask about the kinds of flexibility enabled by overheated cities and the ways in which different kinds and levels of scale clash.

Informality: flexible overheating

It took an anthropologist to discover and theorise the informal sector of the economy properly. Doing fieldwork in open-air markets in Ghana in the early 1970s, Keith Hart was struck by the fact that many of the transactions took place as barter, gifting and other forms of exchange that did not fit into the received models from economics. There were no cash registers, written receipts or tax forms entailed. Should a police officer question the origin of some of the commodities – a bag of shoes smuggled in from Togo, a crate of oranges, a bottle of whisky – he might be paid in kind or cash to keep his counsel. In other words, there was a large and diverse 'underground economy' which went under the radar of the UN and international NGOs, which was not registered in national GDP statistics or accounted for by development agencies. Of course, economists and others had long been aware of the existence of a 'black economy', but Hart (1973) went further by showing not only that a very substantial part of a country's economy, for example the Ghanaian, was 'black' or informal, and that development statistics therefore said little about the actual state of the economy. By connecting the transactions of the informal sector to a broader social analysis, Hart also showcased the strengths of economic anthropology as opposed to economics more narrowly defined by indicating how economic activities are embedded in the social fabric as such and not separate from it. Debts, loans, gifts and favours presuppose and create trust, build or destroy a person's

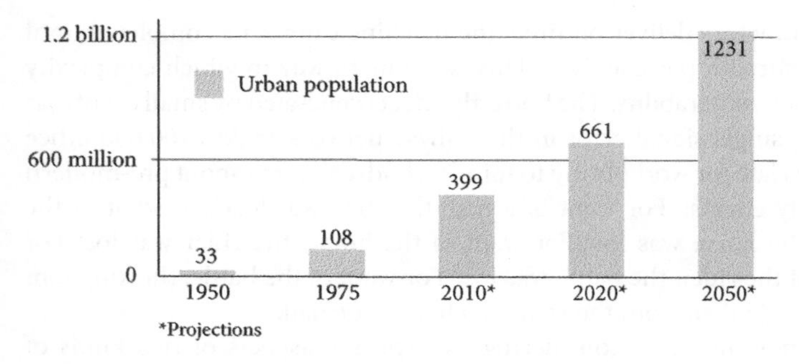

Figure 5.2 Projected urbanisation in Africa, 1950–2050

Source: UN (2014).

honour, create conditions for security and insecurity, and your web of indebtedness tells you where to go and not to go if you need support. The formal sector lacks some of these qualities, at least in principle, in so far as it severs economic activities from social life through an institutional differentiation characteristic of modernity. What Hart showed in his study of the informal sector, was not mainly that Africans were not quite modern yet, but rather that the modern sectorisation of society is artificial and goes against a more fundamental social logic where the economy, like all other dimensions of life, is permeated with moral valuations. In order to function at all, the formal sector also needs an element of informality – interpersonal trust, informal arrangements, the odd Christmas present and other lubricating mechanisms increasing the flexibility of the system.

As argued above, a major problem in the overheating city is the discrepancy between a growing population and a stagnant infrastructure; another problem is the vulnerability resulting from reliance on complex systems for sustenance, where collapse in just a single node may be sufficient for the entire system to grind to a halt. If we consider large-scale modern agribusiness, crucial for the life-support systems of the expanding, overheated megacities, a petroleum shortage would be fatal both to production and distribution, but so would a sudden shortage of phosphates (crucial for chemical fertiliser) or the massive spread of plant disease (such as the Black Sigatoka ailment affecting banana crops across the tropical zone). Comparing monocultural, large-scale plantation-style agriculture with the traditional adaptation of a peasant in a tropical area shows how flexibility tends to be reduced as a result of upscaling, which inevitably results in simplification. The peasant family, producing for its own reproduction as well as a varying surplus for the market, has many small sources of subsistence, meaning that if disease kills the maize, they can eat beans. The mono-crop-based plantation is less resilient and adaptive, since it is doomed if drought, falling prices or disease reduces the value of its only crop. Perhaps the informal sector can be seen as analogous to the peasant adaptation, while the formal sector – with its greater rigidity, reliance on just-in-time delivery and dependence on a steady supply of high-quality energy through the large-scale grid – resembles the inflexible, but highly productive plantation? This is not just a question of academic interest or an experiment in abductive thinking: since flexibility refers to the potential of doing things differently, it matters considerably which forms of flexibility and inflexibility are exhibited in individual lives, communities, societies and kinds of activities at all levels of scale. This is particularly the case with fast growing cities, where flexibility at the systemic level can be very limited, leading to extreme vulnerability. (Just as I am writing this, a news story reports that Bangkok [pop. 7 million] is slowly running out of fresh water, since water consumption is up and the seasonal rains have failed for several consecutive years.)

There is a tendency that increased flexibility in some areas leads to reduced flexibility in others and vice versa, and this is where the informal sector is especially interesting. It could thus be argued that Gutenberg's extraordinary invention led to enhanced flexibility in the transmission of information, but to a loss of flexibility in linguistic variation (it led to the standardisation of dialects) and in locally embedded, experience-based worldviews (knowledge was frozen and externalised); but at another level, there was a flexibility gain in the potential of communicating over a larger area with more people, albeit in a shared, often simplified idiom. Similarly, the car made people living outside of the city spatially more flexible, but less flexible locally. The car pulled them out of the local milieu and deprived them of some of the moral ties that could have been drawn upon in their relationship to their neighbours. The telephone had similar effects.

Nobody knows how many people take part in the informal sector, defined as economic activities which involve transactions in money or kind, but are not recorded officially; but in many countries, informal employment far exceeds that in the formal sector (ILO 2014). Some move between formality and informality depending on the fluctuations in the local labour market. Some have informal jobs in sectors which are strictly speaking formal, such as manufacturing. Others combine informality with food production for subsistence, while yet others see their jobs slide from formality to informality. It is not least in this fluid grey zone between steady employment and improvisational survivalism that the new concept of the precariat (Standing 2011) is useful. Precarious labour is usually formal, but it is uncertain and insecure, temporary and with no promise of permanence. The informal sector grows at a fast pace, along with the growth of the urban areas in the Global South, while precarious labour is a worldwide phenomenon.

The kind of informal economy originally studied by Hart in Ghana is familiar to anthropologists, and it is not confined to the Global South. Even in Norway, probably one of the least informalised economies in the world, there is a thriving market for trafficked goods – cigarettes, alcohol and so on – from other European countries. When soliciting the services of a carpenter or painter, you are sometimes asked, discreetly, whether you want the job done with or without a receipt. To this may be added not only the drug and sex trade, but also flows of various goods and services which are not recorded anywhere, saving the parties involved taxes. However, in these mainly formalised economies, the informal sector remains an exception. The situation is different in economic practices, networks and entire systems which are in their entirety based on informality. That would be the situation in an African market, but also in the vast informal settlements that make up the bulk of many cities. If you are poor, urban and do not live in a welfare state, your material reproduction is, technically, an informal matter by definition. You find ways of getting by. Boys may take a few coins for looking after vehicles in a car park or sell biros to drivers waiting for the traffic light to change. Or they may run errands for the friends of their big brothers, who are petty traders. Women may, for example, engage in hairdressing, grow and market vegetables or marijuana, sell sex or run a market stall with smuggled goods from across the border. Men may, similarly, befriend tourists, serve moonshine in speakeasies, scavenge for food or valuables in rubbish dumps, collect bottles, sell drugs, steal, wash or repair cars. The list goes on. Life in the urban South is largely a form of improvisational survivalism. It hangs on a shoestring, but it is enormously flexible. If there is no water in the tap, you use your networks to get some drinking water; and you can do without a wash for a day or so. If there is no electricity, you'll be fine, since you do not depend on it for your survival. If there are no tourists around, you tend your tiny garden instead.

One of the most fascinating accounts of the informal sector, dubbed 'the world's other economy', is a collection of articles by anthropologists who have studied informal trading across the continents, from the wholesalers (nowadays mostly in China) via hubs such as Dubai and San Diego to small entrepreneurs and consumers (Mathews et al. 2012). Following people, objects and dreams along 'the new Silk Road' (Pliez 2012), delving into the labyrinths of Hong Kong's Chungking Mansions, where African traders buy electronics and mobile phones at a discount (Mathews 2012), or tracing the pirate CD business in Mexico (Aguiar 2012) or Kolkata (Bandyopadhyay 2012), this book gives an alternative image of an overheated world. It is not about the growth in large-scale infrastructure projects, transnational mining or agricultural operations, international climate negotiations or the spread of powerful global chains such as Starbucks or brands like Apple. Yet, the globalisation from below or 'grassroots globalization' (Ribeiro 2006) contributes to making the world more connected, and it positively thrives on overheating. Mobility has become easier and less expensive, goods have become abundant and affordable (especially if they are pirated), and urban growth, where far more people than before can afford to buy a pair of fake Ray-Bans or a fake smartphone, enables thousands of traders to make a profit by crossing borders and negotiating prices with wholesalers.

In a chapter about the trade routes between Yiwu (China) and Cairo, Pliez (2012) segments the Egyptian population by class. The richest 3.5 million, he says, buy original brands and sometimes travel to cities like Istanbul and Dubai as well as shopping in the expensive malls in central Cairo. The middle class, about 15 million, buy less expensive goods, but mainly in the formal economy. In addition, an estimated 40 million people 'comprise the enormous consumer market for cheap clothes, where Asian products ... have replaced low-quality garments made in Egypt' (Pliez 2012: 22). At the very bottom of the social hierarchy are about 40 million Egyptians who are too poor even to consume pirated and smuggled goods.

Although the purchasing power of the working-class segment is far below that of the richest, they are numerous and have money to spare for cheap consumer goods.

Small-scale urban trade in the informal sector is truly part of a global web of connections. In Guangzhou alone, there are at any time enough Africans to support several African restaurants and to earn their quarter the nickname 'Chocolate City' or even 'Little Harlem', and there is a 'Muslim street' in Yiwu frequented by traders from the Middle East.

In Mexican cities, there is a thriving market for CDs and DVDs. The blank discs are smuggled in from China, sometimes via the US, by migrating Chinese entrepreneurs or Mexican business groups (or crime syndicates, if you prefer). The CDs are purchased by people who own multiple-drive PCs (in many cases also imported from China), enabling them to burn as many as 350 CDs a day; and are then sold to the retailers with passable colour copies of their covers. However, the source material – film or music – also has to be obtained, and the small entrepreneurs buy it from people who are able to get originals illegally from the production company, often before their official release. On the other hand, the system is far from borderless. In fact, the very borders that pose challenges to the flow of goods from China (and elsewhere) are simultaneously necessary for the traders to make a profit. A borderless world is not in their interest. When border trade between the US and Mexico was relaxed in the late 1990s, as a result of the NAFTA (North American Free Trade Agreement), informal trading with American goods in Mexico dried up since the commodities could be imported legally. Around the same time, cheap Chinese products began to enter the market, both legally and illegally. A new web of connections was almost immediately established for the import of goods from China.

In many cases, the risk of being intercepted at border crossings, raided by police, fined or imprisoned is inversely proportional to the size of the organisation. Powerful businessmen (or mafiosi) are well connected and capable of paying substantial bribes, unlike the small vendors and CD producers. Yet everyone involved in informal trading runs a risk, from the Chinese counterfeiters to the Mexican scarf vendors. Many find themselves halfway between being inside and outside the law; in markets in the Global North, vendors may pay rent, and even tax, but their goods are as likely as not to be smuggled or illegal copies (or 'replicas', as some traders prefer to call them).

* * *

Is it immoral, Mathews and Vega ask in the introduction to the book, to sell copies of Louis Vuitton bags? 'Perhaps, since it deprives Louis Vuitton of profits on the basis of its design creation, but perhaps not, since almost all of those who buy copies in the poorer areas of the world will never be able to afford the original' (Mathews and Vega 2012: 9). Cheap copies of inferior quality enable them to join the world of consumerism, well aware that the products are not made to last.

Several of the contributors to *Globalization from Below* ask if the informal trade does not create 'a better neoliberalism', that is a free-trade system in which not only powerful companies are able to profit from global consumerism, but where small traders, and equally small producers, can enjoy the same benefits, albeit on a more modest scale? Clearly, by exploiting vacant niches and creating new ones, and by efficiently promoting consumerism among millions of the world's underprivileged, the trade in the informal sector – from stolen cigars sold on the street in Havana to replica Levis – does give consumer opportunities to people who are otherwise deprived, and jobs to millions who would otherwise have been unemployed.

Some mainstream development theorists assume that the informal sector will eventually become formalised. The economist Hernando de Soto (2007) famously argued that formalisation of slum dwellers' shacks would empower them by giving them capital enabling loans and small investments; similarly, the founder of Grameen Bank, Nobel laureate Mohammad Yunus, believes that microcredits will tilt grassroots economies towards the legitimate (Mathews and Vega 2012: 13). Considering the current massive growth in city populations, the continuation of fiscal regimes restricting the transnational flow of goods, and huge social inequalities within as well as between countries, this seems unlikely. Informal trading, just like informal service provision and production in cottage industries and sweatshops, is flexible, resilient and versatile. Capital investments are modest, and the risk is distributed among millions of small businesspeople. When the US market for imports declined, small Mexican vendors turned to China, bringing Chinese rather than American goods across the border at Tijuana in what is called *fayuca hormiga* or 'the ant trade' (Gauthier 2012). If a street peddler is evicted from her preferred spot, she is likely to find a new one, or she might just return. Bankruptcy among informal traders is devastating at the personal level, but has no systemic effects, and it is easier for a small trader to start anew than for a large company. Slums have often been razed to the ground by governments, but they tend to re-emerge. The choice of building materials is flexible, and the labour input required is limited.

Logically, it may seem that while the informal sector depends on the formal sector for its existence, the opposite does not hold. I should like to challenge this view. Considering the extent of social inequality in the booming cities in the South, and most of their citizens' limited access to the formal labour market, it is difficult to see how these cities would have been viable without informality. The producers of the fake iPhones and jeans, be they in the Philippines or Bangladesh, also rely on informality for their livelihood. It has also been argued, from a perspective of pure profitability, that the US economy depends on a steady influx of informal labour (Harris 2002), since it is inexpensive and, by default, exempt from taxation. The informal sector, from trade and retail to production and services, is far more flexible in accommodating new workers and shifting to new areas than the formal economy. Its growth certainly signifies overheating in the cities, even runaway growth; and the growth in transnational petty trade depends on an industrial growth which is far from ecologically sustainable, but seen in the context of the contemporary city, informality fills niches, provides work and satisfies real needs for millions on both sides of the invisible counter. It increases the flexibility and adaptability of the growing urban conurbations in the Global South.

Yet, at the end of the day, a complex society cannot be managed solely through informality. Although a vast number of people engage part- or full-time in an informal economy, it does not create large-scale systems capable of producing a collective surplus. The informal sector does not, by definition, pay for the basic infrastructure of a complex society, from electric grids to schools, roads and police. On the other hand, millions of urbanites today are entirely informal: they live in informal settlements, have informal income, and receive next to nothing in benefits from the state. Moreover, it is not required of people, 'surplus populations', who get by in the informal sector that they provide the basic infrastructure of society. In fact, informality thrives mainly in areas with a growing formal sector. Keep in mind that most of the chapters in *Globalization from Below* deal with goods from China, mainly produced either in the formal sector or relying on it. As Elisabeth Schober points out (personal communication), informal trade often follows in the wake of formal investments. In the Philippines, she says, the influx of foreign direct investments (FDIs) from South Korea has 'brought a number of other Koreans along, who are looking for smaller opportunities because big capital in a way is paving the way for them'. In this way, the formal and informal may be seen as two sides of the same coin, two complementary and mutually constituted forms of urbanity in the overheated city of the early twenty-first century. The way it looks now, both are going to be needed in the foreseeable future. This is not to say that informality does not bring its own problems, and not only for those who lead uncertain lives with no safety net. The image of Bali painted at the outset of the chapter on mobility revealed a society with growing private wealth and a struggling public sector. More people than before can afford cars and motorbikes, but public transport is erratic and unsatisfactory, and the government is unable to improve infrastructures and services at a rhythm matching the growth of human activity. Informality has similar consequences of reducing the relative leverage of the public sector, thereby contributing to its weakening.

The superdiverse city

Both accelerated urban growth and informality are mainly associated with urban sprawl in the Global South. At the same time, accelerated change of a kind currently leading to serious ideological and political frictions also takes place in the Global North. Although the cities grow slowly in absolute numbers, their demographic composition has

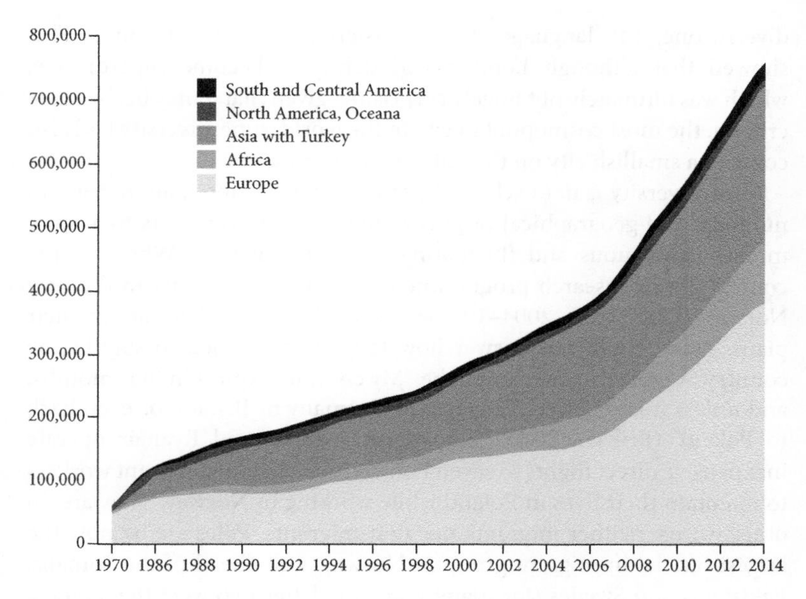

Figure 5.3 Immigrant population in Norway, 1970–2014

Source: Statistics Norway (2015).

changed perceptibly since the mid twentieth century and, as with the other tendencies analysed in this book, the development has shifted gears in the last few decades. I am taking Norway, and Oslo, as my main example here. Berlin, London or Stockholm might have done the job, but Oslo is particularly interesting for several reasons.

I was first introduced to Steven Vertovec's concept 'superdiversity' after a seminar which Vertovec gave at the University of Oslo in 2006. We went out for a drink after his talk, and Vertovec spoke about a situation where 'in the past, migrants came from a few places and went to a few places, whereas they now come from many places and go to many places'. He went on to describe the situation in London, where the largest new immigrant groups had no prior connection to the Commonwealth, tended not to live in particular, delineated parts of the cities, and spoke a bewildering variety of languages – I believe he mentioned the number 300. A year later, he published his now famous article about superdiversity, which outlines an urban situation where diversity had become more diverse than before (Vertovec 2007).

Following this conversation, I went home and checked the latest population statistics for my hometown. It turned out that in just one of the 16 boroughs that make up Oslo, admittedly the most ethnically diverse one, 148 languages were registered at the latest count. This showed that although London had definitely become superdiverse, which was ultimately not hugely surprising, given that it may be, by some criteria, the most cosmopolitan city in the world, superdiversity had also come to a smallish city on the outskirts of Europe.

Superdiversity is not exclusively, or even mainly, about an increase in numbers and geographical origins of migrants; it also refers to a more mobile, ambiguous and fluctuating urban population. When, in the context of our research programme on cultural complexity in the new Norway (CULCOM, 2004–10), we asked Polish workers about their plans, many could not answer how long they planned to stay in the country. A typical answer might be, 'My contract expires in four months, and unless it is renewed, I may go to Germany or Britain, or even back to Poland' (Eriksen 2010). Norwegian Airlines and Ryanair operate inexpensive direct flights to seven Polish cities, enabling migrant workers to maintain their lives in Poland while working in Norway. They are, in other words, neither migrants nor non-migrants. Poles are by now the largest official immigrant group in Norway, well ahead of the Somalis, Pakistanis and Swedes (for many years, the latter two were the largest); but there are also tens of thousands of Poles in the country who have not formally moved to Norway and are unlikely to do so. In a superdiverse city, the structural position of migrants varies considerably. There are the old labour migrants and their children, as well as refugees and their children. But there are also students who hang on after finishing their studies, tourists who somehow forgot to go home after their holiday and found work in the informal sector, foreigners married to Norwegian citizens (and this group also bifurcates into distinct categories – Russian, Thai and Filipina women for Norwegian men, Pakistani men and women for Pakistani-Norwegian spouses, and so on), asylum seekers whose application was turned down and who subsequently went underground, young Swedes who work as shop assistants while saving for their studies or their big tour of South America, German doctors who find work more easily in Norway than at home, and a sprinkling of Dutch migrants who enjoy the spacious layout of the country and the easy cultural compatibility. As Jan Blommaert (2010) has showed, superdiversity creates not only a complex cultural universe, but also a linguistic situation where new varieties of major languages develop at high speed (in countries like his native Belgium, typically Dutch [Flemish], French and English) in order for new arrivals to be able to communicate with each other and with the settled population. Many are temporary; although net immigration into Norway has soared since the late twentieth century, emigration is also considerable.

An older, but no less useful term than superdiversity, Nina Glick Schiller's concept 'transmigration' (Glick Schiller et al. 1995) directs attention to the complementary, but often conflicting social obligations experienced by migrants who may find themselves in a perennially liminal situation. Instead of simply being fully assimilated into the new places they arrive at, as European migrants to the USA might be after a generation or so, transmigrants continue to engage simultaneously with home and host societies through a web of transnational obligations and practices, which significantly complicate old regimes of bounded nation-states. Finally, it should also be kept in mind that superdiverse communities of transmigrants also need to be understood through the lens of citizenship (see Lazar 2014 for contributions from anthropologists), which adds a further dimension to the study of superdiverse communities through its emphasis on state effects of inclusion and exclusion based on the state's concepts of legitimacy and desirability.

The new urban dynamism may be a demographer's nightmare, but it can also be an anthropologist's dream. Oslo is particularly interesting because Norway has become something of an immigrant magnet since the late 1980s. Its immigration regime is not more liberal than that of most Western European countries (it has been less generous than the Swedish one, and the country was widely criticised for its harsh treatment of Syrian refugees in 2016), but many migrants wish to move there because of its booming economy lubricated by oil, and its well-developed welfare state.

In 1995, there were 200,000 immigrants in Norway (which had a total population of 4.5 million, now >5 million). Twenty years later, the figure was above 800,000 (including children of two immigrants), which amounts to more than a 400 per cent increase in 20 years. This development is not unique to Norway but, as a percentage, the immigrant population has grown faster in that country than elsewhere in Northern Europe since the late twentieth century. In 1970, the minority population accounted for less than 2 per cent; in 1995, it had increased to 5 per cent, and by 2015 to 13 per cent.

* * *

Since research on the new forms of multiculturalism and ethnic diversity in complex cities began in the 1960s, questions regarding cosmopolitanism, cultural mixing, withdrawal and ethnic frictions have been raised, explicitly as well as implicitly. Does migration and urban complexity, through increasingly exposing us to each other's lives, lead to enhanced solidarity, tolerance and sympathy with people who are not like 'ourselves' in every respect; or does it instead lead to ferocious counter-reactions in the form of stubborn identity politics – nationalism, religious fundamentalism, racism and so on? This question has, perhaps, a short answer. Accelerated globalisation and its accompanying urban diversity does make it easier for people to communicate and relate to each other across cultural divides, but it also creates tensions between groups that were formerly isolated from each other, and it stimulates among some a need to demarcate uniqueness, separateness and, sometimes, historical rootedness. Strong group identities may serve several purposes – economic, political, existential – in a world otherwise full of movement and turmoil. It may also alienate a minority, or parts of a minority, from mainstream society. Among the neo-nationalist movements in European countries, Jobbik in Hungary and Golden Dawn in Greece may be the most uncompromising, but the Swedish Democrats, the True Finns, the Norwegian Progress Party and UKIP also openly thrive on suspicion towards the stranger. The other side of the coin is withdrawal and entrenchment among some members of ethnic minorities, especially widespread among Muslims. While most members of both majority and minority communities search for a modus vivendi marked by compromise, common values and mutual adjustment, there are those who see no other option than open conflict, sometimes framing it as a clash of civilisations or as the rightful struggle of the true believers against the infidels.

This is all well known. Not all zones or sites in the superdiverse city are harmonious, but neither are they all overheated and tense. In a study of Hackney in London, Susanne Wessendorf (2014) describes a neighbourhood whose residents develop an ability to communicate and socialise in public spaces without foregrounding race, culture and religion. They learn superdiversity skills, the ability to navigate in a terrain where you continuously meet people about whom you know very little. At the opposite extreme, it is also by now well known that small groups, or even individuals, of a conflictual bent are capable of instilling fear and anxiety in very large parts of a population, as well as violence. So if migration and superdiversity in themselves signify overheating – heightened mobility, complexity and friction – the most dramatic forms of overheating are in this respect produced by a few people who cannot bear to 'feel the heat'. Unlike several other European cities, Oslo has not to date experienced violence from Islamist groups, but was the site of a deadly terrorist attack from a right-wing extremist in 2011, when 77 mainly white Norwegians, many of them teenagers, were massacred by a man convinced that his act was necessary for Europe to be able to rid itself of its Muslim immigrants (Bangstad 2014; Eriksen 2014b). In this kind of world, it doesn't take armies or large-scale mobilisations to create very considerable harm. It is the awareness of this possibility, not actual risk, that prompted the term 'risk society' (Beck 1992), and an overheated world is one in which awareness of vulnerability is very much present.

* * *

At a very general level, the superdiverse city, like urbanisation in general, can be seen as an Anthropocene effect. Its relationship to neoliberalism is less obvious, but the increased transnational mobility is clearly connected to the deregulation of markets. On the other hand, as Mathews (2012) has written with regard to informality, the mobility of people is far more constrained than that of capital and (declared) goods across the continents; and it goes without saying that not all migrants are welcome. However, this form of boundary-making is consistent with a neoliberal approach: migration is encouraged under neoliberalism in so far as it is economically profitable, otherwise not. This does not entail that all migrants are profitable, of course; many survive in the informal sector (and do not pay taxes), many are refugees granted asylum in accordance with international treaties and so on.

The economics of migration can be studied as a clash of scale mainly at two levels. First, it has often been shown that migration is beneficial at the level of the global economy since it increases labour participation, consumption and productivity on a global scale. It may nevertheless be unprofitable for certain countries both in the North and the South: In a country like Norway, with its high state expenditure on welfare and other public services, a typical argument is that the current level of immigration does not pay off economically, since immigrants require more state support than they contribute through taxes. Conversely, there are countries in the South where the complaint is that talented and entrepreneurially minded people leave for greener pastures abroad; that is the phenomenon commonly known as 'brain drain'. In other words, what is good for the world economy is not necessarily good for individual countries or local communities.

Second, although life in crowded multi-ethnic neighbourhoods may be stressful for some of the old inhabitants, and although immigration may in some cases not be profitable for the country, living conditions usually improve for the immigrants when they move from one country to another. At their interpersonal or individual level of scale, thus, things get better, and also for their family members who may receive remittances from the migrants; but the situation looks different from another, intermediate level of scale. Controversies over immigration in European countries do to some extent reflect differences in values, but they can also be understood as clashes of cognitive scales: those who oppose immigration emphasise the levels of the community, the native population and sometimes the nation-state, while its supporters emphasise the global situation of inequality with reference to human rights and social justice, and the level of the individual migrant and their family, whose lives usually improve.

Can migration into Northern cities, which leads to a superdiverse situation, fruitfully be seen as a runaway process? At the time of writing, the answer is no. The runaway process is characterised by a lack of thermostats and regulating measures, and thereby ends, since perpetual growth is impossible, in bursting bubbles and sudden collapse. The current immigration regimes in the rich countries are regulated through national legislation, border controls, international treaties and the policing of boundaries, and, although there are loopholes, immigration does not resemble financial markets in terms of growth patterns. If, however, border controls cease to function effectively, climate change makes already difficult lives unbearable, and large proportions of the fast growing Middle Eastern and African populations see opportunities in Europe, we may yet see waves of unstoppable, desperate multitudes in search of a better life. The scenario is not inconceivable.

There are nevertheless indications of runaway processes connected to migration in some rich countries, in the spiralling polarisation between different forms of identity politics, both resulting from migration: extreme nationalists rejecting immigration as unnatural and/or fearing a Muslim takeover, and extreme identity politics directed against the godless infidels by Muslims who see world domination for their religion as an honourable and realistic goal.

The superdiverse city can be extremely flexible in accommodating culturally diverse groups, as research on conviviality and everyday cosmopolitanism has shown (Hall and Josephides 2013). It may nevertheless lose this flexibility at the level of public social life if antagonistic identity politics create a gridlock in arenas where traffic used to flow fairly easily. Nationalists and politicised Muslims tend to blame each other and, especially when confrontations turn violent, they can easily contribute to a polarisation which affects broader population segments as well.

The city is by definition an extraordinarily complex social form built on the premise of a sophisticated division of labour and status differentiation, coordination of several levels of scale from the domestic to those of municipal politics, public services, infrastructure and external relations, and there is no reason to assume that the different levels of scale will grow in consistent or congruent ways. This is a major reason why cities will be a main site for clashing scales and quests for social, ecological and economic sustainability in the years and decades to come. The other, more obvious reason is that most of us already live in cities anyway.

6. Waste

During the first decade of this century, I wrote four books in Norwegian about unintended side-effects of modernity (Eriksen 2001, 2004, 2008, 2011). Only the first, *Tyranny of the Moment*, was published in English. I must have been too lazy, or perhaps too busy, to translate and rewrite the others. Each of them focused on one central paradox or double bind. *Tyranny of the Moment* was about the way in which the recent abundance of information seems to have made us less well informed than before, and how time-saving gadgets, from computers to mobile phones, had only made time scarcer and people more stressed in the rich parts of the world. To this subject I return in the next chapter of this book.

Røtter og føtter – 'Roots and Boots' perhaps (literally 'Roots and Feet') – concerned the paradoxes of identity, and dealt with the duality of tradition (roots) and change (boots). Whereas globalisation makes us more similar in many respects, since we increasingly participate in the same systems of production and consumption, and are more aware of each other than before, people try by all means to be different. The general formula is that the more similar we become, the more different we try to be. However, the more different we try to be, the more similar we become, since everybody follows the same global grammar concerning the expression of cultural difference.

The third book, *Storeulvsyndromet* ('The Zeke Wolf Syndrome'), kicks off from a story in a Donald Duck comic book I recalled from my childhood, where the Big Bad Wolf (Zeke), who has for years tried to catch, cook and eat the three mouthwatering, pink pigs living nearby, finally succeeds in tying them up and dragging them into his shack. Just as the pot is heating up, however, he releases them, realising, in a brief glimpse of enlightenment, that the hunt itself was what made his life meaningful. The book is about happiness and the good life, and argues that what matters most may not be what we think, or rather, what we pretend to think. What matters are the quality of our relationships to other people, fairness and the right to do something challenging and get some recognition by others for our efforts.

The fourth and final book in the series took on the most obvious subject of all, but it took a long time to write, since new source material came my way almost every day. *Søppel* – 'Waste' – is exactly about that very large, smelly and predominant side-effect of modernity, with which this chapter deals.

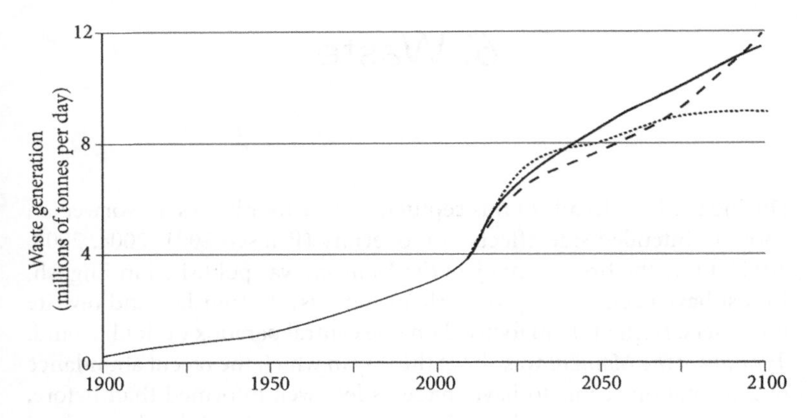

Figure 6.1 Projected global waste production, 1900–2100

Source: Smithsonian Institution (2015).

Waste is the most visible, perceptible and odorous of all the side-effects of accelerated change. It is difficult to delineate and define, since the relationship between waste and value is complicated and shifting, but it is difficult to deny that the amount of waste is growing rapidly worldwide. Waste is the slimy underbelly of consumption, like the stone Antoine Roquentin in Sartre's novel *Nausea* finds on the beach, picks up and examines. Its surface is smooth and dry, but underneath, it is slimy and undefined. Roquentin feels nauseous. Long before eating disorders became indexical of a civilisation that had lost its confidence, Sartre saw what kinds of objects generated nausea and withdrawal. It was the liminal phenomena, the anomalies, that which is neither outside nor inside, but somewhere in-between. Waste is, as Mary Douglas famously said, matter out of place (Douglas 1966). Energy leads to waste in the form of pollution. Mobile people leave waste behind. Cities produce inordinate and growing amounts of waste. The information revolution has led to the production not only of exponentially growing amounts of electronic waste – spam, idle gossip, useless information – but also physical waste, as witnessed in places like Agbogbloshie.

In China, major rivers are being choked by islands of waste disrupting the flow of water. In Pakistan, children develop rare and sometimes fatal diseases due to their line of work, which is to scavenge metals and other marketable things from derelict ships sent there from the North to die. I have already described the swimming experience off the western Balinese coast; and just as recycling is becoming a major industry in the Global North, Southern cities and territories are being poisoned, polluted and turned into wastelands by rapid economic change without an accompanying development of waste management. The global commons, the great oceans, are visibly affected everywhere. The Great Pacific Garbage Patch was discovered only in 1997 by the American Charles Moore, on his way home from Asia in his sailing boat. This floating island, which mostly consists of plastic ground into tiny pieces by the currents, covers roughly the same area as Texas. It is made up of wrappers, plastic bags and other kinds of debris accidentally or deliberately thrown into the sea on both sides of the Pacific, driven to its present location between Hawaii and Oregon by strong ocean currents.

Waste production is by definition a result of affluence and surplus. Egalitarian small-scale societies produce only minimal amounts of waste. Even in complex North European and European societies, waste production was modest a few generations ago, as the figure at the beginning of this chapter suggests. Food remainders were eaten by domestic animals, servants or beggars, and broken things were mended, not discarded. In most Northern countries, there was a whole social class, immortalised by writers like Charles Dickens and Hans Christian Andersen, which made its living by collecting rags, bottles and fat, which they could later sell for a few bob. Scraps of cloth could be used to make fancy quilts. Clothes were either sewn at home or by a tailor; they were expensive and were continuously patched up. Newspapers and catalogues from department stores were used as toilet paper. Moreover, wrapping tended not to be disposable. If some paper, a jute bag or a wooden crate should turn up, it was not difficult to find a use for it, even among the wealthier classes. Many companies delivered flour and animal fodder in sacks printed with beautiful patterns on them, regularly and usefully recycled. The producers competed for customers by making their patterns as beautiful as possible.

The traffic metaphors I used earlier fit well in this domain too. When the world was not yet *trop plein* or overheated, there were limits to the amount of damage humans were capable of inflicting on their surroundings. Locally, the unintended consequences in the form of pollution might be very considerable. The Thames was a stone dead and horribly smelly river in the mid nineteenth century, pastoral nomads had turned the Arabian peninsula into desert much earlier, and in the 1870s, boys could earn a penny or two by shovelling away the rubbish piles from New York streets, enabling the middle classes to arrive at work without smelling of rotting vegetables. Yet, the planet as a whole did well then. It is only now that we are running out of empty places where rubbish can easily be dumped without irreparably damaging a natural environment or a human settlement nearby. In the rich parts of the world, recycling has become a concept sufficiently potent to fuel an election campaign. Today, in the Anthropocene, it is easier to conceptualise the planet as a spaceship than as an endless prairie or steppe; it has become a place where space is scarce and has to be allocated, and where the *matter out of place* concept comes into its own more than ever, since there are fewer and fewer places where this kind of matter might actually be appropriate.

Depending on where you look, the amount of waste can be said both to increase and decrease: there is an indisputable increase since far more is thrown away, from food to clothes and industrial waste; but in some countries, there seems to be a slow but certain decrease in the amount of waste, partly because of large-scale recycling, partly because many rich countries export much of their rubbish to poorer countries.

The world's largest rubbish dump, Fresh Kills on Staten Island, covered an area of more than 20 square kilometres, was over 200 metres high, and was the largest human-made structure on the planet when it was closed down in 2001 (to be re-opened briefly to receive rubble and debris from the 9/11 terrorist attack). Tokyo has long ago filled up its harbour basin, and has for years searched for appropriate spaces to fill with household waste. In my native Norway, we discarded more than twice as much in 2014 as we did in 1994. The bins have become larger, and they are being emptied more frequently than before. And the threshold for defining something as waste has been lowered. A friend, who had bought an inexpensive scanner, told me that it didn't quite work the way it should, so he phoned the store. 'Just bring the receipt, and we'll give you a new one,' they responded. 'But what about the one I bought an hour ago?' he asked. 'Do whatever you feel like. You can throw it in the container for electronic waste outside the shop if you want to.' The cost of having a new scanner produced in China and delivered to a shop in Oslo was less than the act of just turning on the meter for a repair in Norway. Perhaps the defective scanner soon reappears in a container full of electronic waste on its way to China, where the metals and glass are picked out, melted and reconfigured to form part of a new scanner. If so, the scanner enters into an interesting ecological circuit where money is earned and saved at both ends, thanks to the uneven distribution of wealth in the world.

Waste and the global middle class

Many readers will have learned at school that the world's largest human-made structure is the Great Wall of China. This was doubtless the case for a couple of thousand years, but not at the end of the last century. Fresh Kills had far surpassed the Wall in volume when it was closed down in 2001, and it was said that the landfill, like the Great

Wall, was visible from space. (This may be true, but it does depend on where in space you were looking from.) At most, 20 barges arrived daily, each carrying 650 metric tonnes of waste. The residents of Staten Island, mainly white and middle-class, did not exactly rejoice (and would later play a major part in effecting its closure), but they did accustom themselves to daily reminders of the unpalatable side-effects of consumption. Besides, the master urban planner Robert Moses had, before the opening of Fresh Kills, promised the residents parks and new roads as a form of compensation.

Like all landfills dating from the mid twentieth century, Fresh Kills was unsealed. There was no membrane separating the rubbish from the surrounding area. The runoff seeped into and through the mud, turned into untreated sludge and mixed with the brackish water of the Hudson River's estuary, was washed out towards the sea and entered the digestive systems of fish and the cells of algae. Heavy metals sank into the soil. Scrap metal dealers and scavengers sifted through the stinking mass in search of valuables. It is said that those who spend their days at a landfill, whether they are municipal workers or poor gatherers, never quite succeed in washing off the smell. Staten Island morphs into *Satin Island* in Tom McCarthy's eponymous novel (2015), whose narrator, a corporate anthropologist, struggles to make sense of 'the contemporary' with his Lévi-Strauss and his Malinowski on prominent display in his toolkit; drawn towards chaotic traffic scenes and unmanageable oil spills, he ends his search in New York harbour looking across the water towards Staten, or Satin, Island.

Today, New York City's waste is sent away, but it has to go somewhere, so trucks now shuttle back and forth between New York and other states, some as far away as Arizona, chock-full of compressed rubbish. Quite a bit is placed in the poorer states of Virginia and Pennsylvania, where the city has bought land. Quite a lot is also sent out of the country in large container ships. Indeed, large-scale movements of rubbish from rich to poor countries have become common practice. For example, there is an ongoing row over several shiploads of rubbish sent from Canada to the Philippines without the proper permits, at the time of writing slowly and fragrantly decomposing off the wharf in Subic Bay (Elisabeth Schober, personal communication).

You can move waste between continents, but it does not disappear. The chronic problem of getting rid of toxic, or just useless, excess matter does not go away, seen at a global level of scale, just because Americans can persuade the Chinese to store it for them. In 2015, 64 per cent of the household waste produced in the US nevertheless still ended up in landfills – a figure which is far above that of many European countries.

Scientific garbology has a long history, since waste has always been an important, often crucial, data source for archaeology. However, rubbish has caught the interest of other social sciences only more recently, in spite of Mary Douglas' pathbreaking 1966 book, *Purity and Danger*. Her student Michael Thompson wrote a remarkably original study of the dynamic relationship between waste and value entitled *Rubbish Theory* some years later (Thompson 1979), but systematic interest in waste as a cultural category, an intellectual challenge and an environmental problem has reached the social sciences more recently. By the 2010s, however, a good number of anthropologists are studying waste, from Joshua Reno's research in rubbish dumps (Reno 2009) to Tommy Ose's ethnography of food waste (Ose 2016) and Catherine Alexander's comparison between recycling of goods and of people who had formerly been categorised as 'waste material' (ex-prisoners, rehabilitated drug addicts for example; Alexander 2009). The oldest and most ambitious social science project dealing with waste is probably *The Garbage Project*, started by the anthropologist William Rathje in 1973. In the beginning, Rathje and his students mainly studied how the residents of Tucson, Arizona related to household waste, soon discovering that people threw away far more than they were aware of. They later expanded their perspective, carrying out regular excavations at selected landfills across the US. To Rathje's team, the mountains of stinking rubbish came across as a huge treasure trove, an indispensable key to an understanding of the everyday life of Americans in the twentieth century.

When Rathje and his colleagues reached Fresh Kills, they reacted with awe, and possibly a certain feeling of being overwhelmed. Although it had only existed for half a century, the landfill was the highest point on the east coast. The excavations were carried out with a hydraulic drill which cut through car wrecks like butter, and which could reach a depth of 35 metres. The first bucket they pulled up contained scraps of still legible newspapers, which made dating of the layers simple.

The team eventually drilled 14 wells of varying depths at Fresh Kills. All the researchers wore a kind of spacesuit tied to a vehicle. Falling into a well would have meant certain death; a canary in a cage would have suffocated within seconds. Rathje, who had worked with rubbish for 15 years, describes the smell of the first bucket, dated to 1977, as 'penetrating, somewhere between sweet and repellent', roughly the way one might describe the smell of mature cheese (Rathje and Murphy 2001).

The richer you are, the more waste you produce. Americans, who were ahead of everybody else in this respect through most of the twentieth century, still throw away 200 times as much per capita as do Indians. In India, with its booming economy, there are still millions who happily use a cardboard box to fortify a wall or who give a plate of uneaten rice to someone hungrier than themselves. In poor countries, many still scarcely throw away anything. You never know when someone might need that broken iron stove, that piece of wood or those worn gloves. At the same time, countries like India produce a huge and growing amount of waste, industrial as well as domestic. Frequently, those whose surroundings are damaged by the slag heaps of the mining industry or the stink of untreated sewage are themselves very modest contributors to the growing waste problem of their own life-worlds.

Like the energy dilemma, urbanisation and mobility, waste is a rapidly growing phenomenon all over the world; unevenly distributed and with different implications for places and people in different situations, but impossible to ignore anywhere. Beyond the most local level of scale, there are affluent people everywhere, and even the most famished African country has a small rich elite, a struggling but increasingly consumerist middle class, a few factories and often some foreign-owned mines. In other words, the waste problem is a global issue, operating in a multitude of ways, also expressed at the varying levels of scale and, not least, in the clashes between them.

In the last few decades, the waste disposal industry in some countries of the Global North has begun to tackle the central double bind – growth versus sustainability – in highly visible ways. Waste disposal has gone from being a practical problem via a hygienic and sanitary problem of growing significance, to a flagship in municipal environmental policy. Recycling of an increasing number of 'fractions' has become commonplace and, as a result, the landfill has been rendered obsolete in some countries. That which cannot be recycled is converted to electricity through incineration. Against those who are sanguine about the prospects for and assumed consequences of recycling, it has nevertheless been pointed out that one holiday trip from Northern Europe to Thailand is roughly equivalent, in terms of carbon footprints and their offsets, to ten years of diligent recycling.

The ultimate goal of the new ideology of recycling in the Global North is that virtually all kinds of waste should be returned to where they belong; that is, either in nature, in museums or in economic value creation. As I have already indicated, there is nevertheless a hierarchical and transnational dimension to this value creation; the value is often extracted from the rubbish in a country with cheap labour and returned to the rich country in the shape of a product devoid of any trace of the history of its components. Regardless of the global division of labour, the plan is nonetheless that the dump, where poor kids and rats might search for small treasures, should eventually be relegated to the landfill of history.

Living on the landfill

That is to say, in deeply unequal countries, the landfill is still a haven of plenty which feeds thousands of people. I have seen urban slums where the residents are coping reasonably well (nobody starves, everybody has clothes, there is the occasional party), and where the children's toys are made of packaging. Empty matchboxes, spools for thread and cardboard boxes recur. Then you may glance at the houses they inhabit, and discover that they, too, are partly made of packaging. The foundation is built of planks from wooden crates; further up, there may be plywood plates, large sheets of cardboard and the odd advertising poster in aluminium or tin, while the window panes are made of thick, transparent plastic. Those who live in these dwellings, described from a different angle in the chapter on urbanisation, are not on the brink of death. According to the circumstances, they are not doing too badly. As the edited volume *Economies of Recycling* shows (Alexander and Reno 2012), informal cottage industries of varying scale have emerged around the world, not least in the Global South, enabling thousands of otherwise marginal and unemployed people to make a reasonable living recycling other people's waste.

Others are affected by the waste explosion in ways which may be very different. We have all heard about urban scavengers who eke out a living on rubbish dumps, and in a previous chapter I described the poor, north Ghanaian migrants to Accra who sold metal parts from electronic waste in Agbogbloshie. A very different story is that of the Warao, an indigenous people settled along the rivers of the Orinoco delta in eastern Venezuela and western Guyana. Their ethnographer is Christian Sørhaug (Sørhaug 2012). Traditionally, the Warao are fishers and gatherers, with a bit of hunting on the side. They live in pole houses to evade the tides which flood the low-lying land every day, and use canoes for transportation. The word 'Warao' simply means 'those who live along the river's edge'. And a mighty river it is. It was only on Columbus' third voyage to America, when he saw the enormous masses of water flowing through the Orinoco delta, that he realised that he had reached a continent and not merely an archipelago east of India.

Much more recently, the Warao have been increasingly exposed to the modern world, with its ready-made clothes, consumer goods and fast food. For want of a monetary economy, the pleasures of modernity are mostly unattainable. However, just after the turn of the millennium, a new cultural practice was inaugurated among the Warao, one which would bring a broad selection of modern objects to their villages. One day in 2001, a young Warao man Ely and two of his friends returned after an absence of several months, their canoe filled to the brim with cardboard boxes and bags filled with all sorts of things. The villagers were excited. In the coming days, villagers stopped in front of the house where Ely lived, in order to inspect the goods. Many received gifts, mostly clothes salvaged from the landfill.

Ciudad Guayana is located 300 km upstream from the Warao village. It is a rapidly growing industrial city with about a million residents at the latest count. The landfill Cambalache is just outside the city. It takes about a week to paddle from the Warao villages to the landfill, a trip first undertaken by Ely and his friends. This pilgrimage soon became a regular practice among Warao, who plan their trips meticulously and are usually gone for months before they return, laden with treasures.

Everybody who has been there agrees that Cambalache stinks. 'When you are on top of a hill, you may get a few mouthfuls of fresh air,' Sørhaug writes, 'but the moment you enter one of the valleys, you notice that the air is putrid, saturated by slimy matter. I stepped on old diapers, slid on rotting food, and got acute indigestion.' Some Warao have settled permanently in Cambalache. They live in houses made of cardboard and pallets, loosely stabilised with chicken wire and a handful of nails, with a kerosene burner and a latrine. Sørhaug's guide in the dump, a Warao he calls Raimundo, had lived there for eleven years. He owned a large hog which trundled about in an enclosure. Neither of the two were planning to go anywhere.

The landfill is enormous, especially considering that it only caters to about a million people living in a country classified as middle-income. Sometimes municipal workers arrive on excavators and bulldozers to even out bumps and depressions, in order to facilitate the transportation of the waste. The Warao fear these large machines and stay away from them. Afterwards, the workers make new roads through the rubbish. The waste landscape acquires a life of its own.

Not only Warao have their subsistence in Cambalache. Poor Creoles also live there periodically, making a meagre living by selling bottles and scrap metal to middlemen who have installed themselves in modest huts around the dump. The Warao call the Creoles *hotarao*, the people from the hard country, by contrast to themselves, who come from the soft soil typical of the landscape lining the many branches of the river. The *hotarao* sell snacks and other small items at the landfill, and a diverse informal economy unfolds on and around the dump. The Warao tell stories about substantial values that sometimes turn up there – wads of banknotes, gold, diamonds and so on – but add that in order to find them, you have to dig for a long time and be lucky in addition.

Those who live at the landfill approach it in a more systematic way than those who just stay there for a short while. The former systematically collect tin, aluminium and iron which they sell to other Warao, who run small shops through which they distribute the scrap metal to recycling factories in town. In addition, they collect clothes, toys, shoes, electronics and so on, which they often sell to other Warao or have delivered to relatives in the delta. They often find edible things as well. Many tell stories about how they have found half chickens in unadulterated wrapping from a fast food chain, 'and it was still warm'.

The Warao see many parallels between life in the delta and life on the landfill. In both locations, they live in comparable houses. The landscape changes now and then; in the delta owing to flooding, in the landfill because of clean-up operations – and both places smell of decomposing organic matter, albeit of different kinds. They use the same term – 'koera' – to designate both kinds of smell.

The integration of the Warao into the modern world through their engagement with Cambalache seems almost seamless. Yet their current adaptation stands, metonymically, for the hierarchies and exclusions characteristic of an overheated world. Left to scavenge for scraps and leftovers from the affluent, they are the human equivalents of the African dung beetle. The loss of autonomy resulting from their slow shift from hunting and fishing to rubbish collecting means that they are no longer a small-scale society, but an underclass in a large-scale society. If, in a recent past, they were connected to the world through water, fish, game and canoes, their connections are increasingly through cardboard, empty soda bottles, discarded clothes and old transistor radios. In a market society, the Warao, inefficient consumers and poor contributors to GDP, are becoming human waste, excess, a superfluous population.

Meanings of waste

In *Wasted Lives*, Bauman (2004) reflects on the analogies between rubbish and the dual meaning of human waste: bodily secretions and people who are 'matter out of place'. The fateful see-saw of inclusion and exclusion, so endemic to neoliberal globalisation, can be read through the analytical prism of waste, as Bauman shows: waste can be thrown out deliberately or merely ignored. It is rarely talked about unless necessary, but unceremoniously discarded to keep the boundaries clear and the borders crisp. While the extraordinary growth of the sheer amount of waste in the literal sense – rubbish, pollution, smog, slag heaps – is one of the most striking expressions of the Anthropocene, the routine exclusion of humans from secure sociality and meaningful activities, thereby turning them into 'matter out of place', is also an important overheating effect. People who are not needed at the higher scalar levels than their own life-worlds are inefficient producers and consumers; they may be unsuccessful migrants or slum dwellers, Gypsies or 'white trash'. The term 'warehousing', used in the US about the storage of convicted criminals in large penal institutions, is now also used in Israel about the storage of Palestinians in Gaza and the West Bank. The idea of recycling them seems to have been abandoned.

By noting that 'the channels for draining human surplus are blocked', Bauman (2004: 71) helps to create a bridge between this chapter and the previous ones. The bloated cities of the Global South are forced to support far more people than they were initially intended for, ever stricter migration regimes keep 'human waste' outside the borders of the rich countries, and in this overheated situation, not only are huge amounts of literal and metaphoric waste generated, but the excess increasingly has nowhere to go. The refugee crisis in the Mediterranean is a case in point – many refugees can neither return nor settle in Europe, and are thus left in a liminal state – but so is the export of rubbish from North to South, which may soon dry out owing to stricter regulations in the countries, such as China and the Philippines, which have so far received the rubbish of the rich.

The treatment of large groups of people as human waste is far from unknown from history. The very idea that human beings everywhere are endowed with rights and dignity is a recent and comparatively unusual one. In past societies, enemies should be slain or avoided, criminals executed or evicted, heretics persecuted. What is characteristic of the forms of exclusion typical of contemporary state societies is the double bind occurring between neoliberal practices and human rights principles. According to the latter, all human lives have value, but the marketisation of social life implies that activities can be ranked hierarchically according to their economic significance. As a result, those whose activities do not register as economically useful have less value than those who make valuable contributions. In practice, this means that lives lost at sea in the Mediterranean are far more valuable if they are tourists on a sailboat than if they are refugees in a crowded dinghy; the victims of a terrorist attack in a Parisian magazine office far more precious than the victims of a terrorist attack on a Syrian market square. There are thus two double binds associated with the runaway growth of waste: the familiar issue of growth versus sustainability (waste in the literal sense), and the contradiction between universal human rights and the neoliberal grading of human value on an economic scale (waste as useless humans).

In theory, if perhaps not in practice, there is a solution, namely a shift from the logic of spiralling growth that we associate with global neoliberalism, towards a logic of recycling and self-reproducing, sustainable circuits of resources. This holds true not only for waste in its human and non-human forms, but also with respect to other overheating phenomena. A shift towards cyclical reproduction has the potential, in principle, to turn all kinds of runaway processes into sustainable systems capable of reproducing themselves indefinitely. This scenario nevertheless presupposes a shift away from the growth ideology and reliance on fossil fuels.

7. Information Overload

On a trip abroad in the summer of 2015, I picked up a newspaper. Reading a physical newspaper felt almost anachronistic, all the more so given that this was a broadsheet dominated by text rather than images. An article on page 9 caught my attention. Since 2010, it said, the number of photos taken every year had tripled, from 0.35 trillion to a trillion, and was expected to grow to 1.3 trillion by 2017. The estimated proportion of photos taken with a telephone had grown from 40 per cent to 79 per cent in just five years (*International New York Times*, 23 July 2015).

That summer and autumn, world news was dominated by stories about the large number of Syrians fleeing war and persecution and the mixed reactions to their plight in European countries. Some of the more memorable images involved information technology. One photo depicted a young woman just arrived on the beach of a Greek island, in the act of taking a selfie with her mobile phone. Presumably, she would immediately send it, provided her phone had a SIM card that worked in Europe, to relatives at home as evidence that she had made it alive. Another press photo showed a group of Syrian men, huddled together on the floor of Budapest's main railway station, most of them fiddling with their smartphones. In all likelihood, they were communicating with family or friends who were also on the move, or scouring the web for advice about how and where to get asylum in Europe right now.

These images tell a story about accelerated change in two senses: the technology and infrastructure enabling people just arrived in rubber dinghies on Greek islands to communicate instantaneously and transnationally has itself developed very fast, and it is by default a technology of accelerated communication. So, in other words, speedy transnational communication has developed speedily. Indeed, nowhere else has exponential growth been more visible and tangible since the late 1980s than in the domain of information, where transnational connectedness, the sheer amount of available information, the development of new ways of communicating and the inclusion of new groups contribute to the creation of a seamless world of communication and information, where everyday life, probably for billions of people, has been transformed perceptibly in the space of a couple of decades. There are many, partly competing, partly overlapping interpretations of these changes. Some speak simply of 'the pollution of brains'; some see transformations of sociality resulting from the new technologies; some are wary of the

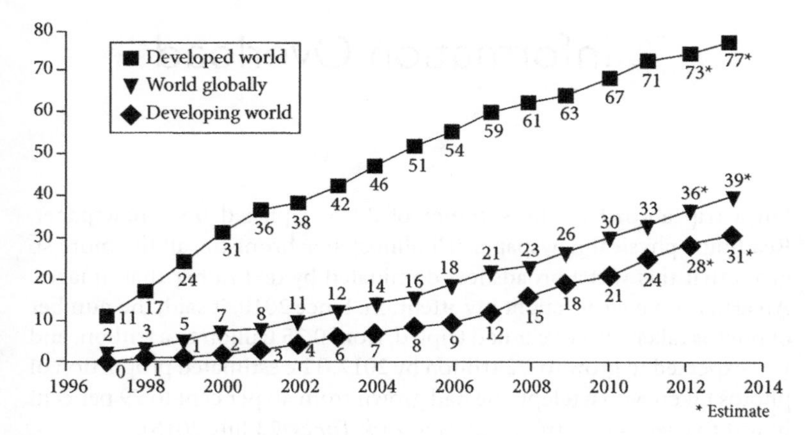

Figure 7.1 Internet users, 1996–2014

Source: ITU (International Telecommunications Union 2015).

potentials of surveillance and control engendered by the internet; while others see the net mainly as a liberating device, making information freer and easier to share than before, as well as paving the way for a myriad of global conversations enhancing humanity's potentials for solidarity, intercultural tolerance and a global consciousness about large-scale contradictions such as inequality, war and climate change.

Being dependent on electronic networks for going about your everyday business makes you vulnerable. In the aftermath of hurricane Sandy in 2012, which destroyed infrastructure along the eastern seaboard in the US, groups of people could be seen huddled outside Starbucks and McDonald's outlets. They had temporarily lost their wifi networks at home, and were using the restaurants' free internet in order to tend to urgent matters. Even a day without connectivity might spell disaster to them.

Anthropologists writing about new information and communication technologies have often been concerned to demonstrate their embeddedness in pre-existing life-worlds and, as a consequence, the diversity and local specificity of apparently uniform technologies (Uimonen 2001; Horst and Miller 2012). In the present context, my main interest lies somewhere else, namely in exploring some of the ways in which information technology is a driving force and an essential element in current global overheating processes. I will argue that the overheating effects of information technology are structurally similar to those of the other substantial areas discussed in this book – fossil fuels, growing cities, increased mobility, waste production – and lead to analogous con-

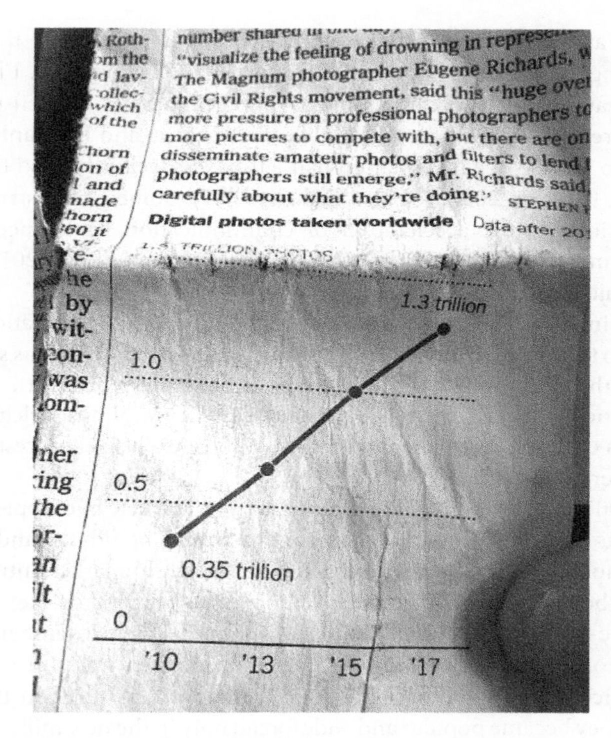

Figure 7.2 Photo of tattered page from *The International New York Times* taken with the author's mobile phone, July 2015

tradictions and paradoxes. Although there is irreducible diversity in the way people around the world access, use or are deprived of information technology, the emphasis here is on the ways in which otherwise very different communities and life-worlds are being affected by a global process; just as climate change affects communities around the world, and the deregulation of the economy changes livelihoods and prospects from Albania to Zimbabwe, instantaneous, scale-free communication is increasingly part of the everyday lives of people everywhere (not all people, mind you), as the image of the Syrian girl on the Greek beach indicates.

The newness of the contemporary world

The speed of recent technological change and the penetration of these technologies into societies around the world has been remarkable. When, in 2001, I wrote *Tyranny of the Moment*, my intention was to demonstrate, analyse and criticise two main unintended consequences of the new technologies (mainly computers, the internet and cellphones). I looked at the way in which so-called time-saving technologies had left people more stressed and with less flexibility than before; and how unlimited access to mostly free information had left us less well informed than in the past (Eriksen 2001; see also Eriksen 2007). Some of the structural implications of the acceleration of communication have since been studied more systematically by others, notably Hartmut Rosa (2013) and Judy Wajcman (2015).

Back in 2001, no more than 10 per cent of the world's population had access to the internet; in the OECD countries, the proportion was slightly over a third. Although some cellphones were equipped with 'WAP' capabilities, enabling them to access rudimentary, almost telegraphic versions of newspapers, the internet was almost exclusively accessed via computers.

In 2001, MP3 players were already on the market, but Apple's first iPod was launched only in September that year. The iPhone and other touchphones ('smartphones') did not enter the world market until 2007. Digital books were still in their infancy; the current market leader among e-book readers, the Kindle, was launched in November 2007. Books could be purchased online, but Amazon.com was still running at a deficit. Although blogs ('weblogs') came into existence in the late 1990s, they became popular and widespread only in the new millennium. According to the largest blogging platform, Wordpress, a new blog is posted on the internet every half a second (Wordpress 2015).

Back in 2001, text messaging was sufficiently marginal not to feature prominently in my book. In 2015, 8 trillion text messages were sent worldwide in a year (Bloomberg 2015). Most of the readers of this book probably send several text messages – a dozen? 30? 50? – on a normal day.

Another significant change since the turn of the millennium concerns the so-called social media, or 'Web 2.0' as this family of services was known at the outset. This is the worldwide web of communication rather than mere information; of sharing, chatting and commenting, the democratising web where everybody gets to have their say, but – the most common objection goes – where the surplus of information makes it difficult to weave a coherent narrative about life, the universe and everything. The video-sharing platform YouTube was inaugurated in 2005, as was Facebook (which had 1.8 billion active users in 2015). Twitter was launched in 2006, Instagram as late as 2010, while Wikipedia came into being as early as 2001, quickly becoming a multilingual, free (or crowdfunded) online encyclopaedia continuously being created by its users.

One obvious result of this massive intensification of the internet is that people worldwide spend more of their time communicating and accessing information online than they did, say, in 2001. (I found the above information online, naturally, and it took me less than a minute.) A no less obvious consequence is the fragmentation of information into ever smaller packets. In the global middle classes, I argued in *Tyranny*, slow time and sustained concentration were becoming scarce resources. Many have since commented on these issues, and there does not seem to be much disagreement. A widely read and discussed article in the *Atlantic Monthly* by Nicholas Carr (2008) asked 'Is Google making us stupid?', responding in the positive. At the outset of the article, Carr confesses that he is no longer capable of reading long stretches of prose continuously: 'Over the past few years I've had an uncomfortable sense that someone, or something, has been tinkering with my brain, remapping the neural circuitry, reprogramming the memory' (2008). His experience, shared by many (including, presumably, many readers of this book), was that his attention span had shrunk and that his ability to absorb and digest information, turning it into knowledge, was now limited to fragments, abstracts, headlines and short texts.

Among the global middle classes, related anxieties are by now common. The novelist Will Self, writing in *The Guardian*, lamented the sad irony of a situation where an art form to which he had dedicated his entire adult life – the literary novel – was now dying before his eyes (Self 2014). The internet activist Eli Pariser, a co-founder of the web-based campaign tool Awaaz, described in detail how search engines 'tailored' and 'personalised' web searches based on people's earlier searches (Pariser 2011). The effect of these services was to entrench and deepen already existing differences between groups. Republicans would get conservative news, Democrats, liberal news. Greens would learn about oil spills when typing the name of an oil company into the search engine, while businessmen would find articles about investment opportunities. Unlike the old-fashioned newspaper, which ideally and often in practice provided a broad and fairly nuanced representation of contentious issues, web engines encourage specialisation and, effectively, polarisation, in Pariser's view.

The cultural pessimism inherent in these perspectives may or may not be exaggerated (come to think of it, I read fewer novels and monographs now than I did a decade or two ago); what is significant is that it has become part and parcel of the critical self-reflection of the global middle classes. Like anxieties over climate change, environmental destruction and mass movements of displaced people, the overheating effects of the information revolution add substance to the view that history lacks a direction.

Digital divides and connections

As in the cases of energy, urbanisation and mobilities, the information revolution reflects, strengthens and transforms global inequalities. English is by far the most widespread language on the internet, there is a concentration of transnational IT companies in the US (all the social media mentioned above were founded by Americans) – and the NSA (National Security Agency) is, as the whistleblower Edward Snowden showed in 2013, far more efficient in spying on its citizens and those of other countries than any other surveillance agency. The English-language version of Wikipedia has two and a half times as many articles as the second-largest one, which is currently, somewhat surprisingly, the Swedish one. The explanation is that a Swedish inventor, Sverker Johansson, has invented a 'bot' which automatically generates up to 10,000 Wikipedia articles a day. In fact, the sixth largest language on Wikipedia is the Filipino language Winaray, which is spoken by just 2.6 million, the vast majority of whom do not have regular internet access (Johansson's wife is from the Philippines).

While most citizens in OECD countries have easy and routine access to the internet, most people in the Global South do not. There are important differences; a far larger proportion of Indians and Brazilians are online than Nigerians and Afghanis. Generally speaking, the poorer a country is in terms of GDP, the less widespread is internet use in that country. However, with the spread of inexpensive mobile phones (not least due to the informal trade networks from China described in an earlier chapter), the digital divide becomes less tidy than one might expect. As early as in 2006, Heather Horst reported that 86 per cent of Jamaicans over the age of 15 owned a cellphone (Horst 2006), and a distinct feature of larger villages in East Africa has for years been the presence of a Vodafone shop. Maasai herders on the savannah are sometimes allowed to charge their phones in game wardens' huts, and the 'cheap calls' described by Vertovec (2004) with reference to phone cards used by non-elite migrants in Europe, enable many Africans to have some access to wireless communication, if not continuously. As previously mentioned, the proportion of sub-Saharan Africans with internet access has grown from about 2 per cent to a quarter in a decade. This figure surely does not imply that 25 per cent of Africans have access to regular, fast and reliable internet services, but that they are occasionally online; and the real figure is bound to be even higher, since smartphones are for lending.

The mobile phone, whether it is used for calling, texting or surfing, extends people's operational networks. Migrants are now capable of staying regularly in touch with their families overseas, parents can monitor their adolescent children more efficiently, and social networking has become more flexible and deterritorialised. Horst (2006) writes that during her fieldwork in a small Jamaican town in 1994, making overseas calls was cumbersome and expensive, while the situation had changed dramatically by the early 2000s, when mobile telecommunications had become ubiquitous and inexpensive. The locals would use the phone for local networking and socialising, staying in contact with family overseas and for coordinating money transfers. Briefly, the introduction of the mobile phone has brought people more closely together socially, in the South just as in the North. The recent and current growth of refugee movements from Asia and Africa to Europe, described in an earlier chapter, cannot be understood independently of the rapid spread of instantaneous communication. Would-be migrants can now be updated momentarily and continuously about migration routes, job opportunities and the whereabouts of their relatives or contact persons in Europe. The virtual mobility engendered by wireless communication enhances the physical mobility of people, and the runaway growth of mobile telephony and internet use witnessed since the 1990s may be paralleled, in the near future, by a similar runaway growth in human mobility. As Horst noted already in 2006, Jamaicans who stayed at home felt relatively deprived when, thanks to the cellphone, they were being updated regularly about life in the faraway countries of their dreams.

Planned attempts to overcome the digital divide at a higher scale than the domestic and interpersonal include development projects intending to help locals in the Global South to take advantage of the new technologies. My student Cathrine Edvardsen studied an attempt by a Norwegian NGO to donate computers for online use in schools in Tanzania and Malawi. In one of the schools she visited, their only computer was locked into a cupboard in the headmaster's office, and was reserved for his use alone (Edvardsen 2006). A complementary study, from village India, showed that the internet cafes set up nearly free of charge for locals to write applications and update themselves on produce prices, were used nearly exclusively by young men for accessing porn sites (Røhnebæk 2006). In other words, the technology in itself is fairly open-ended and does not lend itself to any simple determinism beyond its inherent ability to enhance communication and the retrieval of information. And yet, this latter point is important enough in that it signals the potential of information technology to contribute to societal transformations. There has been much discussion about the role of the cellphone in bringing about the 'Arab Spring' of 2011, but while it is evident that telephones were used actively to coordinate activities, and to film and broadcast the events unfolding in North Africa and the Middle East, there is no evidence that the technology as such played a central part in producing the overheated political situation. What is clear is that instantaneous communication, whether via calls, SMS, internet messaging or other platforms, accelerates social interaction, intensifies communication and amplifies the kinds of acceleration seen in other domains, such as production, consumption and transportation. Although not all parts of the world are equally integrated into this meshwork of accelerated activity, it is being felt nearly everywhere. An image of a Maasai on the savannah comes to mind; the date is January 2008, and he is checking his cellphone while simultaneously herding his flock and looking out for lions, eager for the latest news about the post-election riots in the Rift Valley.

Excess and flexibility

Although global statistics can be effective in demonstrating rapid growth in a number of domains – and have been used to that effect throughout this book – I shall refrain from trying to quantify the growth in 'information' since the early 1990s. The shift from relatively durable, physical mediums of information storage (notably paper) to electronic forms has made quantification difficult and exponential growth inevitable. For those exposed to it, which presumably includes all the readers of this book, this situation leads to a series of familiar problems; the loss of depth to the benefit of breadth, the prioritisation of fragments at the expense of coherence, problems of intellectual orientation, and perhaps an occasional feeling of fatigue owing to information overload. There simply seems to be too much; things fall apart, and the centre cannot hold.

An overheating effect *par excellence*, the excess of information can be seen as analogous to, or even an integral part of, the waste problems discussed in the previous chapter. By default, the increasing availability of free, disembedded and deterritorialised information produces excess; that is, information which is not desired or needed, which prevents the building of a coherent worldview, which overflows and is 'matter out of place'. The anthropologist Fredrik Barth remarked, in 2006, that a major problem for social theory now consisted in the fact that 'our knowledge is growing faster than our ability to give it sharpness and shape' (Gunnar Sørbø, personal communication).

Excessive information can usefully be analysed through Bateson's concept of flexibility seen as uncommitted potential for change. It seems reasonable to assume that a person with access to unlimited information, knowledge, data – the entire Library of Congress, all the newspapers in the world at your fingertips – is far more flexible in his or her way of relating to information, having a wider range of choice at hand, than someone who is restricted to the old media. As many have experienced, this is not the case in practice. In academia, even following a narrow specialisation in detail can be difficult owing to the extreme growth in academic publications. Academic publishing, incidentally, has its own equivalents to the traffic jam. Just as the width and number of roads have been unable to grow at the same speed as the number of motor vehicles, the growth in academic journals has been slower than the growth in academic papers, in some cases leading to queues lasting two years or more for articles already accepted for publication.

But, you may object, all that is required is the ability to select and filter the information you need or want. Well, this is easier said than done. In cognitive theory, a major theoretical issue concerns the way in which knowledge is selected, sifted and organised (see Eriksen 2005). As pointed out in an earlier chapter, the amount of potential knowledge is far greater than that which can be made relevant and useful. The 'uncommitted potential for change' is, accordingly, almost unlimited. However, in order to exploit this flexibility to its fullest, criteria for selection are necessary and, in the overflowing information society, they are more difficult to establish than in a society where information was scarce. Nostalgics may yearn for Kant's study or Plato's academy, chronotopes where anything could happen because nothing in particular happened (that is, times and places where flexibility was high), as opposed to the frantic pace and overfilled brains of the global information age, where flexibility is low because all gaps are being filled.

The amount of information available, but also the speed of its turnover, militates against deep thinking and serious intellectual work. Knowledge, as well as whatever is used to disseminate and store it, becomes obsolete quickly in an overheated world. It is turned into rubbish. Mobile phones rarely last for more than two or three years before they are hopelessly out of date; with computers, it may be five years or so. Some of the obsolete electronic equipment ends up in places like Agbogbloshie, to be recycled in a dangerous, informal, marginally profitable cottage industry in the Global South; but what about the information itself?

Obsolete information does not become valuable in the same way as old books or LP (longplayer) records. It has no tactile qualities; you cannot touch and feel an old electronic text and imagine what its creator might have felt when he wrote it. Electronic books contain exactly the same text as their printed counterparts, but they are more difficult to recall, perhaps because all books become identical on a Kindle. Moreover, the massive amounts of information produced continuously, and available online, sometimes free of charge, sometimes behind a paywall, are not deleted, but accumulate quantitatively, threatening to clog the pipes and fill the information highway with spam, fragments and ultimately irrelevant knowledge.

The situation I sketch is all too familiar to members of the global middle class, and attempts are made to limit the pollution of the virtual space – through spam filters, the 'off' button and so on, but it is no longer considered good form to be offline for more than a day or so. The experience of being drowned in irrelevant information may at times be stronger than the feeling that the physical mountains of waste are overwhelming. Although computers have virtual wastebaskets on their desktops, the hard drives are not actually deleted until they are reformatted. Information waste just accumulates, as in an old-fashioned landfill.

There is a real sense in which information overflow and constant availability overheat people's lives in the global middle classes, thereby reducing their flexibility. As early as 2002, a Danish advertising agency banned email during the core hours, five to six hours a day. The reason was that the constant interruptions made it difficult for the employees to do a proper job. Not long after, a British lawyer was exploring the possibilities of banning employers from phoning their employees after work. The underlying cause for this proposition being made at this time may have been that British employees had, in the space of a short time, lost the pub as a free space. Now that everybody had a mobile phone, they no longer had an excuse for not taking incoming calls at any time, especially if it was their boss (see Hassan and Purser 2007 for relevant approaches to this kind of temporal regime).

The ubiquity of the mobile phone is not a Northern phenomenon. Heather Horst and Daniel Miller begin their book *The Cell Phone* (Horst and Miller 2006) with a story about a bus robbery in Kingston, Jamaica. The robbers wanted the passengers' mobile phones, but reacted angrily when only 26 phones were collected, since there were 29 passengers. They did not find it likely that any of the passengers in the executive bus did not carry a cellphone – and rightly so, as the three remaining phones were duly handed over following a brief, threatening gesture with a gun.

The mobile phone has become the centre of gravity in the world of instantaneous communication. Most photos are now taken with telephones; in addition, it is a music listening device and a game console, an address book and a calendar. You can read books on it and stay updated on email and Facebook. The mobile telephony described by Miller and Horst in 2006 has long been superseded by a far more diverse and versatile technology with which users interact in a different way.

The ubiquity of an indeterminate amount of information not only speeds up life, fills the gaps and intensifies networks; it also creates difficulties in sorting material and determining its relevance. After the terrorist attack on the United States in September 2001, it soon became clear that the authorities had been tipped off about some young men of Middle Eastern origin who were taking flying lessons, but seemed uninterested in learning how to land the plane. Why did they not take action? Because the CIA and FBI received thousands of tips about suspicious people and activities every day, and it was impossible in practice to distinguish the useful leads from the ones that belonged to the dustbin.

Excess of information may well be regarded as a kind of waste. It pollutes not only minds, but also time, filling the gaps and turning slowness into a scarce resource. In this major sense, information reduces flexibility. In Bateson's words, 'flexibility is to specialization as entropy is to negentropy' (Bateson 1972: 505). Negentropy, in this context, refers to total order, specialisation to path-dependence and the strict delimitation of different activities and processes, each interlocking perfectly with the others, like the pieces of a jigsaw puzzle. Maintaining flexibility in a system as a whole, Bateson argues, 'depends upon keeping many of its variables in the middle of their tolerable limits' (1972: 502). Explaining the term further, he speaks of flexibility as elbowroom in his description, referred to in Chapter 2, of a man on a high-wire using his arms to maintain the stability of his body. In a situation where time lags are theoretically absent, and where communication has accelerated to a lightning speed, the gaps – the space where you can flap your arms – is reduced. The contemporary world of information is like an overgrown pond where the algae – necessary and ecologically functional at the outset – have grown uncontrollably and exponentially, making the pond uninhabitable for fish and amphibians. Or perhaps it is best likened to the traffic situation in São Paulo, where a technology aimed at increasing spatial flexibility (motorised vehicles) ends up reducing it (through frequent gridlock).

Connections

There are pattern resemblances, empirical connections and reinforcing mechanisms linking the informatisation of society to other overheating phenomena. As already noted, the migration from Asia and Africa to

Europe, and from Latin America to North America, has been transformed in only a few years by mobile communication. Connections between people on each side of the border are easier and faster, facilitating decision making about mobility, perforating and relativising the border, and reinforcing tendencies to 'superdiversity', that is, the diversification of diversity whereby more people move more often between countries and within them, sometimes forming communities based on language, origins or religion, sometimes not. Electronic communication has also facilitated money transfers through remittances, social control and the transmission of knowledge. For example, among religious specialists in Pakistan, a lucrative business consists in giving lessons in the Qu'ran on Skype to the children of migrants in Western Europe (Aarset 2014). And, as Horst concludes a discussion of the mobile phone in small-town Jamaica: '[t]he availability and ownership of mobile phones has in many ways collapsed the distance between Jamaicans at home and abroad due to their ability to create a sense of involvement in each other's everyday lives' (Horst 2006: 156).

While mobile telephones have for years been crucial for coordinating and organising migration to Europe, the smartphone, introduced in 2007 and dominating the global market since around 2010, broadens the intensity and potentials of instantaneous communication. Migrants and the entrepreneurs who earn good money by helping them out ('human traffickers') can now navigate and coordinate themselves with GPS and photos taken with the phone, in addition to texting, discussing on social media and – still – ringing each other up. The availability of inexpensive smartphones in many African and Asian cities is itself contingent on fast economic growth in China. The informal electronics traders, encountered in an earlier chapter, who travel back and forth between China and cities in Africa and Asia, depend on inexpensive flights, which are now widely available owing to the fast growth in air traffic. Their numbers have also been augmented by the rapid urbanisation in the Global South, whereby millions have been driven off their land, or left it more or less voluntarily, as a result of privatisation and large-scale agrobusiness, mining, population growth and possibly climate change. In many cases, both traders and their customers used to be rural and are now urban.

Urbanisation and population growth enable more people to engage in informal trade since there are more potential traders and far more potential customers; these processes also lead more people to migrate or wish to do so, since the cities are crammed and local opportunities few, temporary and precarious.

The growth of the Chinese economy (and that of other Asian countries) relies on a steady supply of minerals and affordable energy, and the production of mobile phones involves the extraction of a range of minerals, some of them rare. In other words, the extractive industries which have contributed to urbanisation in Africa and industrialisation in China are also integrated in the production of the mobile phones facilitating mobility and communication among the newly urbanised Africans and Asians. Likewise, electricity is just as important for the production of smartphones as it is for their everyday workability.

Finally, in addition to producing massive amounts of digital rubbish, the smartphones are likely to end their days on a rubbish heap, such as Agbogbloshie, where they are teased apart by youngsters in search of the valuable bits.

In these ways, and many others, we see how the global information revolution is empirically and not just metaphorically connected to the other forms of overheating considered in this book – urbanisation and mobility, mining and energy, as well as waste production in all its forms. A complementary approach to the question of connections consists in considering the pattern resemblances between different overheating tendencies. For example, all overheating phenomena discussed can be described as runaway processes, that is, fast growth processes with no inbuilt thermostat or conductor. Information growth is the most striking example of all, since there are no natural limits to it. The potential amount of information in the world is infinite, unlike the case with for example, coal production, the size of cities or the number of tourists on a beach, where natural limits are reached sooner or later. Yet the sense in which the 'global information highway' (the net) is being congested by masses of useless information where searching for the nuggets of gold is like looking for a needle in a haystack, resembles other forms of congestion, where overheating – acceleration – is punctuated by standstill. The traffic jam is the quintessential image, that is, a situation where flexibility is reduced to nil because all vacant spaces are filled. A world in which humans use available resources at maximum speed and intensity is one with severely reduced flexibility both in terms of current livelihood and possible futures. I have argued that we are heading towards such a world, but have also showed how enhanced flexibility may crop up in unexpected places, such as the informal sector and small-scale renewable energy production. Similarly, the information technology that creates path dependency, addictive disorders and virtual gridlocks can also be a tool for scaling down and increasing flexibility. Decentralised services mediated by online connectivity, such as Uber for informal taxi services and Airbnb for temporary accommodation, are sometimes described as 'the future of capitalism' or even of 'postcapitalism' (Mason 2015). Again, the parallel to other overheating phenomena is evident. Rifkin (2011) speaks of a 'Wikipedia of energy' when presenting his idea for a decentralised electricity grid fuelled by you and me, each possessing solar panels and connected to a larger system unmediated by a centralising corporation. Both Mason and Rifkin see small-scale networks fuelled by solar energy and connected by information technology as a counterpoint and alternative to the sluggish, standardising, destructive and centralised corporate interests that currently rule the world. The most obvious objection to such societal models is that they do not specify who, or what, should be responsible for maintaining roads and universities.

These examples show not only that overheating can be synonymous with runaway processes and that flexibility is a scarce resource, but also that clashes of scale are the key sites of conflict in a globally integrated, neoliberal capitalist world. Each family of overheating effects has its own double binds, its own flexibility budgets, crises of reproduction and treadmill spirals leading to runaway processes, but what holds the scaffolding together in my analysis is scale, specifically the conviction that clashing scales illuminate contemporary overheated globalisation, reveal its contradictions and suggest their possible overcoming.

8. Clashing Scales: Understanding Overheating

Allow me briefly to sum up the discussion so far before taking it a step further.

- The contemporary era can rightly be called the Anthropocene, since the human footprint is visible everywhere owing to an exponential increase of human activities, premised on population growth and technological innovations fuelled by non-renewable energy sources and leading to short-term environmental degradation and long-term climate change.
- The currently hegemonic ideological regime worldwide, with important regional variations, encourages marketisation and the deregulation of markets whenever it benefits the most powerful actors. Usually labelled neoliberalism, this ideology tends to translate political issues into economic or managerial ones, thereby avoiding debate about fundamental values, social justice and conditions for long-term human well-being.
- Anthropocene neoliberalism is characterised by runaway processes in a number of separate, but interrelated domains; that is, growth without a mechanism for deciding the upper limit, or to stick to the 'overheating' metaphor, heating with no thermostat. Energy use, urban expansion and population growth, tourism and migratory waves, waste production and its direct effects on local environments are typical runaway processes, as is the phenomenal growth in internet use since 1990 and in transnational trade. Yet the most clear-cut example of runaway neoliberalism is arguably financialisation, where trade in fictitious commodities (money) and the irregular but frequent bursting of bubbles ensures the continued volatility of the global system. Treadmill competition at different scalar levels fuels runaway processes and demonstrates that they have neither upper limits nor a stated objective.
- The central double bind, or unresolvable dilemma, in this era is that obtaining between economic growth and ecological sustainability. Growth precludes sustainability and vice versa, yet politicians and global organisations claim that they favour both.

Other double binds characteristic of the global system can also be identified, such as the tension between the universalism of human rights and the particularism evident in the *de facto* grading of human lives, and possibly the problem of class-based politics versus green politics.

- Upscaling is integral to globalisation, while downscaling is proposed by many of its critics. Although the 'globalisation from below' exemplified in the discussions of migration, the informal economy and small-scale transnational trading is also very widespread, processes at a large, or expanded, scale lead to far more encompassing and consequential changes. Scale can be conceptualised in terms of space, social organisation, cognitive worlds and temporal horizons, and in all areas, scales clash more powerfully and effectively in this overheated world than before. Some of the most familiar clashes of scales take place when a local community is being overrun by large-scale interests, when short-term concerns take precedence over long-term survival, or when local jobs are prioritised at the expense of environmental sustainability.

Upscaling effects

The contemporary world is not accurately described by the term 'clash of civilisations'. The 'clash of civilisations' notion, promulgated by the late Samuel Huntington (1996), sees profound cultural differences as a driving force and major cause of conflicts, which Huntington predicts will erupt along 'faultlines' between the civilisations. Few if any of the conflicts witnessed in the world today run along civilisational faultlines. It is true that many armed conflicts today involve Muslims who position themselves against Western hegemony, but much of the fighting takes place among Muslims, and insurgents such as Daesh ('IS') and Boko Haram can scarcely be seen as representatives of 'Muslim civilisation'. My contention is that the concept of clashing scales is more instructive, versatile and useful in making sense of the frictions and tensions resulting from global neoliberalism, the fossil fuel hegemony and its accompanying environmental problems.

The effects of upscaling have been noted many times in earlier chapters, but let me try to be more succinct here. Scaling up means enlarging something in order to gain some benefit or other. The classic modern version in politics is nationalism (Gellner 1983), whereby the relevant systemic boundaries of life-worlds expand through the effective incorporation of communities into nation-states. The key to success for a nation-state lies in its ability to create congruence between the political scale and the cognitive scale of the inhabitants, ensuring their identification with the imagined community of the nation.

In the currently overheated world, the most consequential instances of scaling up may be those taking place in the economy, where large corporations oust small actors. A telling image from my fieldwork in Australia is this: in January 2014, a new branch of the Australia-wide Coffee Club chain opened on Goondoon Street, the business hub of Gladstone. Across the street, a family-run greasy spoon serving hearty breakfasts and strong, homebrewed coffee had done brisk business for many years. Shortly after the inauguration of the Coffee Club cafe, the 'For Sale' sign appeared in the window of the older cafe. In her book about the tourist explosion, Becker describes a scene from a city recently opened up for mass tourism like this:

> Then he walked out on the street with us and pointed out the stores that had been pushed out by the high-fashion stores on his street. 'That first alley on the left – it used to have a butcher, a florist and a bread shop. All disappeared. It's only the international fashion people who can afford the rents.' (Becker 2013: 84)

Scaling up creates economies of scale. Small businesses are thereby outcompeted, but another general effect of upscaling, which also reduces local flexibility, needs mentioning. Let me take the plantation as an example. The monoculture represented in the plantation, as opposed to the versatility of the peasant, entails simplification on a structural level, standardisation of the workers' skills. Productivity increases, but each worker is unable to make the plantation work on his own, unlike the peasant with his plot and animals. By becoming a wage worker, the peasant may gain a pecuniary advantage, but loses autonomy and flexibility; he is now dependent on the plantation for his livelihood. Plantations also, naturally, reduce biodiversity and thereby ecological flexibility. On the other hand, one should not see the small-scale farm as a panacea. It will not feed the world and, as Robert Pijpers points out (personal communication), if your farm is on infertile soil in West Africa, your family is left with food scarcity every rainy season, you would probably prefer an alienating factory job in town to the uncertainties of small-scale peasant agriculture.

The standardising, simplifying logic of the plantation, described by Mintz (1953) as the prototype of the factory, can be recognised wherever there is industrial development, but it is far more ubiquitous in a globally interconnected era when capital investments travel more easily than earlier. You may just think of the way in which your word processor and presentation software simultaneously shape the way you work and facilitate connections with other people. An almost exact parallel to the Microsoft software, which is still hegemonic among all kinds of writers and presenters around the world, could be the standardisation of the shipping container and its accompanying infrastructure (Levinson 2006). Almost akin to the implementation of a metric system for measurements, the introduction of fixed proportions in shipping containers, beginning in the early 1960s, has transformed world trade to a degree few are aware of. In large ports worldwide, the unruly and often unionised stevedores and dockers are a relic of the past, since goods are now moved from railway carriages or trucks to ships by cranes operated by a few, often well-paid, men. Regarding the organisation of labour, the shift from labour-intensive to mechanised work in the ports parallels the shift, described in an earlier chapter, from labour-intensive coal mining to mechanised oil drilling. But for the standardised shipping containers, which vary in colour and company logos, but not in size and proportions, to conquer the world, railways, trucks, ships and the very layout of ports had to be modified, in many cases built anew. By this point, the world of commodity trade has locked itself to an industry standard, and those who were unable to play by the rules of the Red Queen's treadmill principles were doomed to oblivion, such as the port of Liverpool, which failed to reconfigure itself, partly owing to the strength of the dockers' union, and was sent into a downward spiral towards the end of the 1960s. (A counter-example is provided by Erem and Durrenberger [2008], who tell how a dockers' union in South Carolina, after containerisation, succeeded in scaling up their struggle, eventually enabling dockers across the world to strengthen their position.)

Apart from reconfiguring labour and infrastructure, the transition towards large-scale standardisation in the shape of container ships reduced transportation costs dramatically. Consequently, it now makes economic sense for a toy shop in Indiana to get its Barbie dolls from China, where the labour costs are far lower, than from a local producer. This would not have been the case before the container ship.

Such large-scale, standardised systems for communication, transportation, consumption and production form the infrastructure of the overheated world. The clashes of scale are evident. The dockers' unions in port cities like Liverpool operated on a local scale and were overruled by processes taking place at a transnational scale; toy factories in the Midwest, similarly, were outcompeted by large-scale operations overseas able to ship their goods anywhere in the world at a low price.

Clashes of scale in the economy – the domains of production, distribution and consumption – may be good for the 'world economy' (an abstraction with little relevance for people's lives), but not for those who are directly affected by them. In a study of banana farmers in the Caribbean island of Dominica, 'Overheating' researcher Frida Aamnes (2015) explored local responses to the deregulation of the European market for bananas. Until the 1990s, Dominica, along with other states in the Eastern Caribbean, had for decades after independence been guaranteed access to the UK and Europe and minimum prices for their banana exports under the Lomé I and II agreements. When this preferential treatment ceased, a gradual process throughout the 1990s, the small-scale Dominican banana farmers, mostly just working their own tiny plots, found themselves in a situation where they had to compete with bananas produced on a large scale in mainland Central America. It was impossible to produce bananas at a competitive price on the small Dominican plots. Local politicians tended to blame the farmers, accusing them of laziness; while NGOs active in the island encouraged them to link up with Fairtrade to tap into a niche market for organic bananas produced under socially responsible conditions. The farmers in Aamnes' study tended to opt for flexibility as opposed to the large-scale monoculture of the large plantations and the path dependency implied by membership in Fairtrade (which places many conditions on its producers). Many continued to grow bananas, but supplemented their income in a variety of ways, almost like independent peasants who grow many crops in small quantities and keep a few animals in addition; some ran little establishments where they sold food and drink from their homes, some sold marijuana, some rented out a room, while others might make some money through fishing, driving, roadwork and so on. Their response, in other words, did not consist in trying to scale up, nor to hook up to an alternative large-scale system (Fairtrade), but instead finding ways of surviving while retaining their autonomy. In their case, this presupposed an economy which was legible locally and had a foundation in local practices of production, distribution and consumption, while – inevitably – incorporated in nationwide, regional and transnational systems to varying degrees. Although the Dominican banana farmers miss the old days of quotas and preferential arrangements, they have opted for autonomy rather than dependency on a large-scale organisation over which they have no means of exerting control. What is good for a world committed to economic growth as its central prerogative, namely maximum deregulation and effective exploitation of comparative advantages, may not be good for people in particular localities, who not only need to survive economically, but also demand a degree of autonomy.

Scaling up, finally, can be used to weaken the bargaining power of those who are adversely affected by change. About half an hour's drive west of Gladstone, the East End mine has produced limestone for a nearby cement factory since the 1970s. Farmers in the Mount Larcom area near the mine have long been wary of it since large quantities of water must be pumped out before the limestone can be extracted. They have formed an association and claimed compensation for the loss of groundwater entailed by the mining operation (Eriksen forthcoming b; Lucke 2013). In the past, the local farmer Alec Lucke says (Lucke 2013), they could go and see the manager and sometimes come to an agreement with him. He would then be a local who knew the farmers and their needs well. Today, following an amalgamation, the head office of the mining and cement company is located near Brisbane. When the farmers' association files its complaints, the manager of the mine is no longer a local but a hired professional from overseas, who explains that he cannot help them and that they have to contact the head office. Apart from the practical difficulties of actually achieving a rapport with the head office, my informants explain, the response they get is likely to be a standard letter, detached from the small-scale circumstances of 100 farmers in Central Queensland and their tangible life-worlds. The responsibility for the depletion of the water table has been shifted to a higher scale, making it far more difficult to hold those in power accountable than in the past. This critique summarises, in a nutshell, a main consequence of upscaling, namely the removal of decision making to a higher level of scale, thereby making it difficult to exert influence locally.

A very common complaint in this day and age is the allegation that the locality contributes more to the higher levels of scale than it gets in return. In industrial Gladstone, if people were asked about the significance of their contribution to the Queensland economy, some might comment, a tad sourly, that yes, they made a huge contribution to the economy, not of Queensland but south-eastern Queensland (the densely urbanised area around Brisbane).

Another, perhaps less obvious clash of scales can be seen in environmental politics. Environmentalists in Central Queensland, whose concerns are with specific, local issues such as industrial pollution of the air and sea, often criticise Greenpeace and other green movements operating on a large scale for not making appropriate connections between the large and the small scale. As mentioned in an earlier chapter, one said to me, after commenting on how Greenpeace somehow never visited the city or engaged with the community: 'They are world champions at saving the world, and they're pretty good at saving the

Great Barrier Reef as well, but they don't care about us who are living in this so-called industrial hell.'

Clashes of scale do not only make it difficult for locals to exert an influence on decisions affecting their livelihood. The feeling of being overrun by large-scale economic forces, or of being ignored by large-scale political institutions and NGOs, is also a much publicised source of resentment and often leads to the formulation of locally based alternatives.

Scaling a discussion up one or several levels may be a way of avoiding dealing with the tangible and concrete, as with the case of the environmentalists not engaging with local communities. It may in fact be easier to describe ways of changing the world than ways of changing the practices of people living in a particular place, just as pushing responsibility upwards one notch can in practice prevent a direct confrontation and reinforce power asymmetries. One can easily imagine a marital quarrel where she accuses him of being sloppy and uncouth, while his response is that 'This, darling, is a very gendered discourse characteristic of middle-class Protestant relationships of the kind described by some contemporary social theorists as "pure relationships" at this time.' Needless to say, there is a potentially destructive clash of scales between her complaints concerning his shaving and dressing habits and his contextualisation of her complaint in a generic discourse.

In the opening pages of *The Hitchhiker's Guide to the Galaxy*, Douglas Adams (1979) reveals a profound understanding of the clashes of scales typical of multiscalar systems. Arthur Dent, an unassuming Englishman, is given to understand that Earth is about to be demolished by the Vogons, who plan to build an intergalactic highway through the orbit of our planet. When he exclaims that they cannot do it, Dent is told that the Earthlings have had plenty of time – several thousand years in fact – to submit their complaints and comments, but have not made use of this right, so they are themselves to blame for the destruction of their home. Adams makes an explicit parallel with the local tribulations Dent is going through in connection with the planned demolition of his house; in my reading of the story, it is also an accurate allegory of the relationship of indigenous peoples to the state, or for that matter any small-scale entity being overruled, or overrun, by a large-scale entity. This asymmetrical relationship is much studied by anthropologists and, during the current acceleration of resource extraction, these conflicts – from northern Canada to Australia and Amazonia – often concern mineral rights. 'Overheating' researcher Margrethe Steinert (2015) describes the mounting external, large-scale pressure on the Asháninka, a small ethnic group in lowland Peru, showing that villagers are far from sanguine about their prospects: 'Colono invaders, mining, and oil and gas companies are seen as a threatening counterpart. They are "powerful others" viewed as stronger than themselves and with the intention to steal their land,' she writes (Steinert 2015: 51), adding that the locals have no effective means of negotiating with the government in Lima and the resource companies. The government is nearly as distant as Alpha Centauri to the Earthlings in Adams' narrative.

Blaming and scale

Kierkegaard famously said that life could only be lived forwards, but had to be understood backwards. Similarly, we may say today that while life can only be lived locally, it has to be understood globally. As I have indicated, clashing scales are expressed in very visible and tangible ways in the economy and politics, but they may be no less important and consequential in the cognitive domain, where the world of your experience and reflections thereon may fail to dovetail with the general, perhaps statistical knowledge you get from other sources. When knowledge regimes at different levels of scale clash, people have to choose between them; either you give pride of place to what the experts or media tell you, or you trust your own observations. When the two kinds of knowledge seem to contradict each other, most of us would trust our own experiences rather than the lofty figures of the experts and attribute blame accordingly. So when a Peruvian farmer perceives the rain patterns to be abnormally unstable and irregular, he might connect this to the climate change that NGOs and governmental agencies are speaking about. At the same time, the national meteorological institute might attribute the poor rainfall to normal variation. In addition to blaming climate change, however, he might also scale down and engage with traditional knowledge and ritual practices in the hope that the rain might return (Astrid Stensrud, personal communication).

The problem is that of interpreting a changing world and positioning it on a moral template of trust, blame and responsibility. During the global financial crisis from 2007 onwards, there was a decline in production and income in the mining areas of Sierra Leone and, when asked what they thought was going on, Robert Pijpers' informants there might shrug and say, 'It's the global' (Pijpers, personal communication). What exactly they mean by this statement is uncertain; the point is that they have notions of higher-level systems affecting their local life-worlds, and do not assume that these large-scale systems can be influenced directly, although they may be held responsible for changes affecting the locals. One of the main sources of tension, conflict and discontent in an overheated world is the awareness that local life is being influenced in important ways by remote instances which are only dimly known, if at all. In Anthony Giddens' well-known discussion of trust in abstract systems (Giddens 1990), a considerable degree of predictability is presupposed; for example, money nearly always works in a predictable way, and you know what to expect from traffic lights. In a situation where there is a huge gap between the local and the abstract scales, however, mutual knowledge is generally patchy and the distribution of power deeply asymmetrical. As Giddens correctly says, ontological security in complex societies hinges on trust in abstract systems which individuals at the grassroots level have few ways of influencing.

Although the topic of trust was introduced by Simmel in his *Philosophy of Money* (Simmel 1990 [1900]), the social science literature on trust remains patchy and sprawling. It is nevertheless important for an investigation of overheating and scale. Owing to rapid societal changes, many question the enduring stability and legitimacy of the structures and kinds of social relationship that were capable of producing trust in the past. The standard modernist narrative about 'the expanding circle' and the shift from interpersonal trust to trust in institutions (Gellner 1988; Fukuyama 1995; Ridley 1996) or 'disembedded', abstract phenomena remains influential, but it has increasingly been questioned by research which looks specifically at variations in levels of trust within contemporary complex societies, revealing that people in one and the same society may differ radically as to what or who they trust and distrust (see for example Rothstein 2000). In fast-changing, multiscalar societies, a unilinear account is not very credible; trust in large-scale systems is unlikely if they neither deliver on their promises nor produce knowledge which comes across as true and relevant. You are then likely either to place your trust in an alternative large-scale knowledge regime, or to revert just to trusting people you know personally and your own first-hand experience, blaming the processes on a higher scale for everything that goes wrong. (The routine invocation of neoliberalism as a shorthand description of all kinds of exploitation, ruthless environmental degradation and short-sighted profit-seeking shows that social scientists are no more immune to this temptation to simplify and push responsibility upwards, than everybody else.)

Since abstract knowledge is bound to be contested in a complex, changing society, different knowledge regimes compete for legitimacy and influence. Whereas the most obvious conflict between knowledges is that obtaining between abstract (or 'expert') knowledge and the knowledge of experience, clashes occur not only between cognitive scales but also within the same level of scale. For example, during a lengthy and bitter controversy about the effects of dredging in Gladstone harbour, unfolding from c. 2011 to 2014, both experiential knowledges and expert knowledges clashed. Briefly, some Gladstonites insisted that fish and mudcrabs had become diseased owing to dredging, while others claimed that the situation was neither better nor worse than it had been years ago. Similarly, some experts (hired by the Ports Corporation) argued that the water quality in the harbour was acceptable, while others (committed to environmentalism) insisted that it was pretty toxic (Eriksen forthcoming a). So it is not merely a question of the universalist, abstract, general knowledge of the bird's eye perspective clashing with (or crashing into) the minuscule, localised, particular but accurate gaze of the participant on the ground; there are also clashes within each level of abstraction.

Reconciling a cognitive world based on observation and experience with one based on abstractions is never easy. Lévi-Strauss (1966 [1962]) contrasted the *bricoleur* with the *ingénieur*; someone who creates new ideas with the materials to hand versus someone who begins with an equation or a geometrical principle. Goody (1977) contrasts the context-dependent, embedded world of orality with the more decontextualised, disembedded world of writing, using the contrast as a prism through which to see a broad range of cultural variations. Many others have worked with various dichotomies depicting contrasting levels of cognitive scale (see Berreman 1978 for a useful list); but the realities for most people in today's world is not either–or, but rather both–and. My everyday consumer practices and beliefs have to be reconcilable with what I read and learn about global climate change. Or, my knowledge of my Indonesian *kampung* traditions has to be made compatible with the universalist religion called Islam. Or perhaps, as a Peruvian farmer, my understanding of the relationship between labour and wealth must be related to what I learn about my government's policy and the functioning of the global markets.

Sometimes, the levels of scale cannot be connected at all. Perhaps the neighbouring village has been bombed to smithereens by Western bombs; what should I then think about the same governments' solemn talk about universal human rights? And if the drought in my region, which has led to a food crisis and a miserable life for most of us, is a result of global climate change, it is difficult to call those responsible to account, although optimistic campaigns in favour of climate justice are carried out. Who can I blame, and what can I do, if my life-world is altered dramatically because of climate change? If climate change is to blame, political action becomes far more difficult (who should I then blame, and where should I go?) than if the culprit is a local company. Shifting blame up to higher and more abstract scales thereby enables corporations and governments to behave more ruthlessly than they might otherwise do, since they can divest themselves of responsibility by referring to for example 'the global market' or 'global climate change'.

When cognitive scales clash, people may find ways of smoothing over the contradictions. An environmentally conscious person in the global middle class may justify flying by stating that 'the plane would have departed anyway'. Sliding almost unnoticed from the personal to the macro scale, he fails to see, or refuses to admit, that every passenger is in the same structural position as himself, and that the large scale of the airline service is simply the sum of the small-scale acts by himself and a couple of hundred others, neither more nor less.

Blaming 'the system' is a common local response to overheating effects, whether that system is conceptualised at the level of the region, the state or 'the global'. However, we should look more closely at the different modes of blaming operating when scales clash under conditions of overheating. There may be something universal about forms of blaming as well as the specifics associated with 'blaming the global'* In an important essay on risk and blame, Mary Douglas (1992) distinguishes between three forms of blaming when, for example, a woman dies in the kind of African village society with which she was familiar – accompanied by distinctive reactions from society.

First, the woman might have been to blame herself: she could have offended the ancestors or broken a taboo. In this case, the act of redress from the community would be one of purification and expiation, meant to ensure the future obedience of rules and norms.

Second, the blame could be attributed to an individual adversary or a competitor within the community, in which case the reaction might be that of compensation according to a tit-for-tat logic of justice.

Finally, the cause of the woman's death might be located outside the community: it might be someone who did not respect local norms and who could represent a subversive, disruptive force. It would usually, but not necessarily, be an outsider, but it could also be a traitor. Douglas mentions communal punishment and demands for compensation as typical reactions, but she might have added war or cessation of relations, depending on the perceived gravity of the death inflicted.

It is perfectly possible, and could be instructive, to identify similar modes of blaming in contemporary societies: When something

* The following section is based on the first part of Eriksen (2014b).

is seen to go wrong, it could ultimately be the victim's own fault, it could be the result of conflict, contradiction or competition within society, or it could be the work of insidious outside forces, often seen as some version of 'the global' or its permutations (for example mass immigration, US imperialism, Chinese expansion and so on). Douglas may have underestimated the degree of disagreement and conflicting perspectives present in the traditional societies she wrote about; it is nevertheless beyond doubt that in the complex societies of the early twenty-first century, several modes of blaming are at work simultaneously. Indeed, much of the competition for scarce resources, as well as the most significant ideological conflicts, can be understood in relation to differences between modes of blaming and their relationship to different levels of scale. Surely, modes of blaming (or the attribution of guilt or responsibility) often appear in ambiguous and confusing ways in contemporary societies, characterised as they are by proliferating risks – environmental, economic, social, cultural – and widespread insecurity about the future. As I noted in the first chapter, the belief in progress, crucial to theories of social evolution and an ideological linchpin of capitalism, has been weakened perceptibly almost everywhere in the last decades, largely as a result of the continuous onset of unpredictable crises which have been produced, in the final instance, by the neoliberal world system itself, but which nevertheless inspire a whole range of different modes of blaming. Douglas states: 'There are communities, barely earning the name, which are *not organized at all*: here blame goes in all directions, unpredictably' (Douglas 1992: 6, my italics). The situation in contemporary, complex societies may perhaps be described as intermediate between this, arguably a borderline case, and that of the cohesive moral community where there is agreement about whom to blame and what to do.

To what extent can Douglas's three ideal-typical modes of blaming be seen as relevant in the context of contemporary overheating?

Who is to blame for mass unemployment and the encroaching informalisation of labour? As noted earlier, in some of the smaller banana-growing islands in the Caribbean, there has been a tendency for certain politicians to attribute the weak recent performance of the industry to the laziness of the banana farmers. The latter may blame neoliberalism, as the quotas and guaranteed minimum prices for bananas from the Eastern Caribbean in the EU have successively been abandoned, with dismal consequences for small producers, whose competitive disadvantage vis-à-vis the large-scale banana plantations in countries like Costa Rica already constitutes a serious handicap. The two modes of blaming reveal two opposing social ontologies, one which blames the victim and individualises social processes, and one which points to systemic factors at a higher level of scale, which are harder to identify since they cannot be observed directly. They correspond to two of Douglas's modes of blaming, the missing one being that which attributes blame to an internal competitor or adversary. However, the recent failure of banana exports may also be attributed to the politicians, who have favoured other economic activities and failed to support the banana industry sufficiently for it to remain competitive.

The economic crisis affecting many European countries, which have had to cope with large-scale unemployment and uncertain prospects, has also been explained in a number of ways. Seen from the position of the new precariat – the informalised labour force, which lives in a state of perennial insecurity – three typical explanations are invoked: First, the fact that a person does not have permanent employment could be said to be his own fault. He may have inadequate qualifications, he may have made professional mistakes in earlier jobs, or he has failed to market his skills in a sufficiently confident and convincing way. (There are no important changes in the labour market.) Second, the situation may be blamed on unwanted competition from within, for example foreign workers who are prepared to work under worse conditions than yourself. Blame can thereby be seen as an implication of a zero-sum game rather than an intentional act. Third, the deterioration in working conditions may be attributed to systemic failure, such as deregulation and a generalised neoliberalism which successfully evades social responsibilities, general incompetence and dishonesty in the political classes, or the slow loss of any comparative advantage that European economies might have had, compared to the growth economies, especially of Asia.

So far, Douglas's simple typology of modes of blaming appears to work well in a complex society, just as it does in a simpler one, although in the multiscalar kind of societies in which most contemporaries live, all forms of blaming are present simultaneously. We nevertheless need to add an important dimension, which is not included in Douglas's model: standard accounts of modernity emphasise a movement from the small scale and tangible to the large scale and abstract as a constitutive element of the transition from a traditional to a modern society. Trust, notably, is now expected to be invested in abstract entities such as laws and state institutions, disembodied science and context-independent knowledge. Morality, similarly, is expected to be based on universal principles, not social relations. The critiques of this kind of dichotomous thinking are familiar and relevant, and the point is not that trust and blame in contemporary societies are necessarily associated with abstract, person-independent forces and structures, but that there is a widespread cultural assumption among the hegemonic classes that it should be so.

Does this hold for modes of blaming as well? Is there, in general, a stronger tendency to blame abstract entities in contemporary, overheated societies than in the cooler societies of yesteryear? It would hardly be well-advised to try to answer this question unequivocally: societies where the structures of trust are based chiefly on face-to-face contact, kinship and personal familiarity exist, but large-scale, experience-distant religion often enters into the mode of trust – and therefore also that of blame – in decisive ways. Likewise, in societies where the inhabitants learn to trust institutions and principles rather than individuals, people remain socially embedded in interpersonal relationships, and individuals are frequently blamed for what could equally well be seen as structural failure. In an important sense, however, it may arguably be said that institutions take the place of religion in many modern contexts.

Following the catastrophic earthquake in Lisbon in 1755, at the height of the French Enlightenment, reactions ranged from the religious to the sociological (Shrady 2009). The Jesuit priest Gabriel Malagrida published a book in the year after the earthquake, where he argued that it was God's punishment for the many sins committed by Lisbon's inhabitants. Voltaire wrote a poem about the earthquake where the inherent argument was that this kind of meaningless disaster reveals the cruelty and amorality of nature, reminding humanity that we are on our own and cannot trust or blame God when things go wrong. He would later elaborate this perspective in his famous philosophical novella *Candide*. Rousseau, in a long letter to Voltaire, disagreed. He pointed out that although the earthquake was not humanly induced as such, precautionary measures might have mitigated its disastrous effects – a third of Lisbon's population perished. For example, Rousseau suggested, a more scattered form of settlement as well as a kind of early warning system might have helped save many lives. These three attributions of blame correspond to a great extent with Douglas' types: Malagrida blamed the people themselves and their sinful behaviour; Voltaire blamed an external foe, that is the mindless and amoral forces of nature; while Rousseau suggested that internal arrangements in society itself could at least partly be blamed for the consequences of the disaster. Given the scale of the disaster, it stands to reason that nobody seemed to blame individuals. The diversity of the responses to the earthquake, and the different accounts of its consequences, bear testimony to a plurality of competing worldviews, and bear some resemblance to the reactions in Western Europe to the South-East Asian tsunami of 2004.

The current economic crisis in Europe is similarly accounted for by competing and sometimes contradictory worldviews. Blame is attributed to certain persons within the crisis-ridden societies, who are depicted as enemies of the people – Greek civil servants, German bankers and so on – or outside the societies in question; consider the widespread demonisation of the German chancellor Angela Merkel, not least in Greece. There have been situations where people even blame themselves when the system they have trusted fails. In the early 1990s, Sweden went through a severe economic crisis which led to the devaluation of the Swedish krona, substantial unemployment and a *de facto* decline in wages. At the time, few blamed their own political and economic elites, foreign capitalists or immigrants offering cheap labour. Instead, the general identification with the system was so strong that many Swedes felt that they themselves had failed, since no clear boundary was drawn between state and society, people and elites.

Modes of blaming relating to climate change reveal a similar pattern, which again corresponds roughly to Mary Douglas's tripartite division. Many members of the Western middle classes tend to place the blame on themselves, or their neighbours, for their unsustainable lifestyles. Others blame their own elites for not acting upon extant scientific knowledge. Yet others blame external forces – technology, neoliberal ideology, world capitalism – which come across as nihilistic as Voltaire's godless nature, but which are nevertheless unintended aggregate consequences of human agency for which we are, collectively, ultimately, responsible. Again others deny the reality of climate change, and blame powerful international organisations, governments and scientists for keeping the truth from the public eye.

In general, the more large-scale, abstract and distant whatever, or whoever, is deemed responsible for misfortune, the more difficult it is to act upon one's assumptions about who or what to blame, as several earlier examples in this chapter indicate. Therefore, it can often be socially and psychologically necessary to personalise blame, to reduce the scale and/ or complexity of a phenomenon in order to make it manageable – finding someone to blame, someone to punish and a relevant course of action in the midst of a situation which is *de facto* too complex to deal with properly. As is well known from the classic anthropological functionalist accounts of witchcraft (Evans-Pritchard 1983 [1937]; Nadel 1952), siphoning blame off to a weak or vulnerable agent can efficiently deflect attention from underlying structural conflicts. Conversely, attributing blame to the world system can also deflect attention from local conflicts and responsibilities.

Scaling up, down and sideways

Finding the correct addressee for a social critique is not straightforward in a multiscalar world where the levels can be difficult to distinguish. Pitching the critique or complaint at the appropriate level of scale is therefore difficult, and discourse about overheating effects tends to slide up and down the scales. In rich-country public debates about climate change, the politicians typically scale down to the level of the individual citizens, admonishing them to be more ecologically responsible (naturally without thereby consuming and producing less), while green activists and NGOs scale up by blaming the politicians for not having implemented sufficiently severe laws, fiscal regimes, trade regulations and so on, which would have changed the rules in such a way as to make it rational and reasonable for society's members to move in a more sustainable direction.

Responses to perceived overheating effects usually entail one of three ways of scaling the problems: scaling up, scaling down or scaling sideways.

Scaling up a problem refers to the belief that it should be handled at a higher systemic level. Since individuals, local communities and even states tend to pursue their self-interest, the reasoning goes, supranational arrangements need to be put into place to force actors to collaborate and act for the benefit of the system as a whole, not just their own bit of it. Typical examples of upscaling are international climate summits expected to end in binding treaties, trade agreements like the ones negotiated by the World Trade Organization (WTO), international laws guaranteeing workers' rights or the much-discussed offset system in transnational environmental politics enabling the polluters to pay non-polluters for not doing things they would otherwise have done (such as cutting trees). The assumption here is that higher-order coordination can contribute to the implementation of global norms at the lower levels of scale. In evolutionary theory, 'major transitions' (Smith and Szathmáry 1995) take place when higher-level coordination is capable of quelling tendencies towards destructive competition at the lower levels, creating complementarities rather than competition and enabling higher-order phenomena (such as complex organisms) and lower-order phenomena (such as cells or bacteria) to rely on each other. Attempts have been made recently to apply this biological mechanism to human social organisation. Using Norway as an example of a society with minimal tension between the low and high levels of scale, the biologists David Sloan Wilson and Dag O. Hessen (2014) argue that the logic of the village, based on interpersonal trust and social control, can be expanded and made functional at the level of a multiscalar, complex society.

The most obvious objection to this view is that the village logic is nonscalable: there is an upper limit to the number of persons who can form a *Gemeinschaft* based on reciprocity and trust, and even a relatively small (compared to other nation-states) and ethnically homogeneous (87 per cent ethnic Norwegians) society like Norway relies on large-scale institutions, legislation and sanctions to ensure conformity to norms. But this objection does not really affect the argument, only the village metaphor, which is unfortunate. Provided that the systemic levels where coordination takes place are perceived as legitimate, the model outlined by Wilson and Hessen is largely congruent with the notion of a 'great transition' in evolution. However, it may still turn out to be misleading. First, countries like Norway cannot serve as models for the rest of the world, since they are untypical. Second, shifting from the nation-state level to the planetary level may be unrealistic. Third, it should be kept in mind that the high level of prosperity, trust and equality enjoyed by Norway is financed by a growth economy currently based on oil and gas.

Projects aiming to scale up activities in order to solve problems generated as unintended, cumulative side-effects of lower-level activities cannot, all the same, be discarded as impossible. There has been a series of 'major transitions' in cultural history, massive urbanisation being the latest and currently perhaps most noticeable, whereby people have quickly accustomed themselves to living peacefully and often prospering among strangers, in anonymous settings. The great systems of thought and/or belief originating in 'the Axial age', about 2500 years ago, expanded the cognitive scale, and ideally the moral community, hugely, from village to imagined community of believers. Writing and, more recently, printing, similarly enabled identification with abstract communities on an unprecedented scale. The story of the state is more complicated, since the standardising large-scale logic of states tends to rely on coercion (see Scott 1998 for an influential account with recent case studies), although there are examples of states and other large-scale political entities which enjoy a high level of legitimacy. The question may perhaps therefore be rephrased: under what circumstances are large-scale organisations seen as legitimate by actors whose life-worlds are anchored in small-scale systems? What does it take to create synergies and cooperation rather than conflict and clashes between scales?

Scaling up politically, ultimately to the global level, is based on the conviction that the problems facing humanity in an overheated, increasingly interconnected world cannot be addressed efficiently by communities or states alone. Scaling down, the opposite kind of response, is instead based on the view that abstract entities are unable to build the kind of trust necessary for equality and sustainability to be possible; that they are inimical to cultural, social and biological diversity, disempowering at the local level, homogenising and standardising. There is a long and diverse tradition in political theory and activism exhorting the virtues of the local, manageable and accountable, as opposed to the faceless, insensitive and alienating forces of large-scale organisation. In the heyday of the green countercultures, E.F. Schumacher's *Small Is Beautiful* (Schumacher 1973) was a symbol of a particular way of thinking, which placed a strong moral evaluation on the phenomena I have spoken of as clashes of scale. According to this ideological persuasion, small scale was in principle superior to large scale since it ensured moral accountability, facilitated equality and was conducive to ecological responsibility in ways that were far more difficult to achieve at an abstract level of scale.

In anthropological research, similarly, there is a large literature, accumulated in the course of over a century, which describes small-scale societies with strong egalitarian values and *de facto* ecologically sustainable livelihoods. They are often seen implicitly, and sometimes explicitly, as being morally superior to hierarchical, complex, modern, unequal and ecologically unsustainable societies. This tradition of cultural criticism, which can be traced back at least to Rousseau, sees hierarchies and alienation largely as a product of the great transitions of cultural history, beginning with the slow shift to agricultural practices, the growth of cities, the early state, writing and early imperialism. There is real merit in this approach, witnessed in influential articles like Sahlins' 'Notes on the original affluent society' (Sahlins 1968), monographs like Clastres' *Society Against the State* (Clastres 1987 [1974]) and entire intellectual empires such as Lévi-Straussian structuralism. In line with this tradition, a recent book such as Joy Hendry's *Science and Sustainability* (Hendry 2014) convincingly argues that contemporary people living in complex societies have lost ecological insights that were taken for granted in small-scale societies whose members engaged with their environment in a more intensive and reciprocal way. Yet global modernity cannot expect small-scale traditional societies to solve its own contradictions. An interconnected world of 7 billion cannot conceivably be decomposed into small-scale, autonomous entities. This is why small-scale alternatives can supplement large-scale organisation and mitigate some of its harmful effects, but they cannot replace the coordinating, higher levels of scale. As a response to the Greek economic crisis, local economic practices of barter and cooperative systems of production and distribution at the local level have been revived (Rakopoulos 2014), and these practices are both ecologically sound and enable people to reproduce themselves materially and socially outside the formal economy. Yet, such 'human economies' do not contribute much to the maintenance of hospitals and highways, nor do they satisfy people's desires for iGadgets and motorbikes.

Scaling sideways represents a third kind of response, and it is even more diverse in its forms and less straightforward to describe than the others, since it is activated at many different levels of scale. Scales are typically nested – Barcelona's city flag next to Catalonia's next to Spain's next to Europe's next to the UN flag – whereas sideways scaling does not necessarily change the systemic level, only the mode of organisation. Castells' notion of the 'network society' (Castells 1996) captures the potential of scaling sideways, but without developing it conceptually. This nevertheless takes place continuously in this interwoven world, but tends to be understood through the cognitive template of nested scales. The early indigenous movement in the 1970s is a telling example. Sámi leaders in the far north of Scandinavia were concerned about a planned hydroelectric plant which would dam the Alta River and thereby interfere with the seasonal reindeer migrations. Pleading upwards to the Norwegian state had little effect, but connecting with First Nations in Canada facing comparable issues helped generate international attention around their cause. Eventually, the plight of the Sámi became a news story in *Newsweek*, which means that it was scaled up, but not in a linear way, rather via an incipient network of stateless peoples in comparable situations of powerlessness.

The WCIP (World Council of Indigenous Peoples) and similar worldwide organisations based on shared interests or ideological persuasions, rather than assumed shared origins or a shared space, have developed ways of scaling large problems sideways rather than upwards. This may be so because the formal power holders lack legitimacy or credibility (for example, a hostile state or a profit-seeking corporation), or simply owing to perceived shared interests or persuasions across geographic boundaries. For diasporas, sideways scaling is necessary to maintain contact with the country of origin, and to fulfil moral or contractual obligations towards relatives or associates in the homeland. Instead of being engaged in domestic politics where you live, you engage in long-distance nationalism (Anderson 1992) or something analogous in a place far away.

The social movements that emerged in the wake of WTO summits around the turn of the millennium (Maeckelbergh 2009) typically sidestep the level of formal domestic politics in order to forge strong organisations which may influence nation-states without being part of them. Similarly, political Islam may be seen as a family of deterritorialised alternatives to political identities based on the nested hierarchies of the nation-state.

Sideways scaling can thus operate on various systemic levels. It can be global in its ambitions and have millions of supporters, as in the global environmental or Islamist movements, or tiny, as in networks of support and reciprocity involving kin living in different countries, where remittances contribute to building the economy not in the country in which the wage worker lives, but in the country of his or her significant others. When sideways scaling clashes with other sets of activities, it is not necessarily a matter of clashing levels, but of clashing principles of organisation. Typically, Sharia may clash with secular law, indigenous ideology and rights claims with national politics, informal trading networks with formal distribution of goods. In these cases, it is not always easy to discern which network of activities is located at a higher and lower scale, respectively. Informality may take on many large-scale characteristics (vast distances, an intricate division of labour, central accumulation of surplus and so on), while the scope of the formal sector may, conversely, be restricted by, for example, municipal jurisdiction or national borders.

The big story about the overheated world is one of large scale impinging on and dominating small scale. In the early stages of my fieldwork on industrial expansion in Australia, I jotted down a general – trivial, perhaps, but useful as a rule of thumb – principle on the basis of a handful of cases, namely that when big money meets small money, the big money wins. All other things being equal, economies of scale are more profitable than small-scale enterprises; and large-scale religions such as varieties of Christianity or Islam are far better organised and resourceful than small-scale oral religions. This does not mean that the latter are simply obliterated – small traditions may coexist with large traditions for centuries or even millennia – but that they are marginalised. In a fundamental sense, the dialectics of globalisation concern the tension, not between 'the global and the local', but between the abstract and formal, and the tangible and informal, the universal and the specific, the disembedded and the embedded. It is in this context that the possibilities of decentralising large-scale activities become especially interesting. Trawick and Hornborg (2015) suggest organising large-scale modern societies along the lines of large irrigation cooperatives. Rifkin's (2011) proposal to decentralise energy production and distribution and thereby turn it into a collective project rather than a hierarchical one, comparing a decentralised solar energy grid to Wikipedia, is especially interesting in this context.

One significant aspect of overheating, which is brought to bear on all the empirical areas outlined in this book (and many others), is intensified competition; indeed, runaway processes may be glossed simply as processes of schismogenetic competition with no ceiling or regulating thermostat. One of Bateson's (1972) main examples of a schismogenetic process was the armaments race between the US and the Soviet Union, which eventually resulted in the production of absurd numbers of warheads on both sides. In the contemporary global economy, competition between nation-states or corporations may lead to comparable results. The ceiling exists – the world's physical resources are finite – but from within the logic of economic growth, it is made invisible. It is therefore incumbent upon us to ask whether contemporary civilisation is an Irish elk, or whether it will eventually turn out to be just a harmless peacock's tail.

If we consider human survival through the lens of ecological sustainability, it is necessary to distinguish between renewable and non-renewable resources. That which is renewable can be sold and bought, negotiated and relinquished for a while, since it can be recovered. That which is non-renewable must, accordingly, be guarded, nursed and protected. Throughout human history, until very recently, nature has been perceived as unproblematically renewable. It 'strikes back' at culture, which has to protect itself from the forces of nature. The great economists of the early modern period – Smith, Ricardo, Marx – saw natural resources as inexhaustible. An exception was Marx's contemporary William Jevons, who predicted that industrial growth would exhaust coal supplies. He was concerned with the conditions for development, not ecological side-effects, which were at the time noticed only locally. It is only in the last few decades that nature has increasingly been seen as weak and vulnerable in the face of aggressive and expansive human projects, and in need of the protection of well-intentioned human beings. Thus nature conservationism was born as a cultural project. Fossil fuels are non-renewable unless you take a very long view (several million years), but so is phosphorus (a key ingredient in fertiliser). Perhaps, as some anthropologists have suggested, identity is that which cannot be sold and bought; a non-renewable resource, an inalienable possession, that without which your past, present and future lose their significance (Harrison 2000). Anna Tsing (2012) has described this dimension of local life as 'nonscalable'; it cannot be mass produced, and if it is universalised, it may end up either being simplified beyond recognition or – more interestingly – function as a Trojan horse thwarting the universalising logic from the inside. Following a similar logic, many religious believers hold that simplification is what happened with Islam when a particular, mainly Saudi version of that religion was exported across the Muslim world; its local embeddedness is eradicated, and the religion becomes a set of inflexible, rigid, imposed rules and principles.

The possibility of a global conversation

This short introduction to overheating has documented accelerated change in a number of interrelated domains. The most fundamental driving force, enabling fast population growth, mobility and urbanisation, has been the fast growth in energy production since the outset of the nineteenth century. I have shown how the increasing predominance of large-scale systems creates numerous clashes of scale where the local level of social organisation or cognition repeatedly conflicts with the uniformisation and standardisation, the alienation and domination of large scale. In analysing these clashes of scale resulting from runaway processes of growth and change, I have discussed degrees and forms of flexibility in different settings and at different scalar levels, in the awareness that flexibility at the systemic level is generally reduced when energy consumption increases. An underlying question has been to what extent ecological sustainability is compatible with large-scale, globalised social and economic organisation.

Some commentators are optimists. Whereas Wilson and Hessen (2014) argue that the equitable and sustainable 'village model' of reciprocity and resource management may be extended to a higher systemic level, Richerson (2014) adds that four conditions for this higher-level organisation to be legitimate are meritocracy, tolerance, respect and justice. Positive feedback from the higher levels of scale reinforces the lower-level processes, justifying taxation and contributing to the reproduction of the system as a whole. Along similar lines, Trawick and Hornborg (2015) argue that large-scale economic systems may be organised similarly to collectively managed irrigation systems. All participants benefit from the long-term sustainability of the overall system and act accordingly, favouring a collective, global maximum over an individual, local maximum.

In the contemporary world, the allegories about the peacock's tail and the Irish elk seem more immediately relevant than stories about long-term global sustainability. At the time of writing, there is nothing to indicate that growth is slowing down, or that a massive shift towards renewable energy is imminent. However, one important overheating effect, which does not deplete material resources but could instead make a contribution towards rescuing them, is the expansion of cognitive scale. Just as the Filipino container ship, the Mexican maquiladora, the Chilean grape plantation and the Chinese factory producing Barbie's hair shrink the world of production, distribution and consumption, so do the ever denser deterritorialised networks of communication shrink the world of communication and discussion. Clashes of scale are no less visible here than in the economy, the local or particular clashing with universalist claims and large-scale overviews, but the potential of a global dialogue inherent in electronic communication technology should not be underestimated.

In a sense, this brings us, full circle, back to the beginning. I began this book with a meditation on Lévi-Strauss's lament for a lost world and anxieties over the globalisation of modernity, and argued that our overheated world of accelerated change should be at the centre of academic inquiry into the contemporary. Rather than yearn for purity and boundaries, we must accept impurities, mixing, clashing worlds and cultural universes that change at different speeds and which are rarely in sync even with themselves. Allow me therefore to end with a brief reflection on the conditions and possibilities of a global dialogue, within which it may be possible to slide up and down the scales without losing sight of the global level. For as the history of the last 200 years has made clear, what is profitable at a local level may be catastrophic when projected to a global level; and what makes sense globally may be disastrous locally.

A main premise for a global conversation, apart from a globally integrated economy, shared long-term interests in dealing with climate change and social injustice, and technology facilitating global communication, is a continuous process of cultural hybridisation or creolisation. Cultures are not becoming the same everywhere, but they develop, in unprecedented ways, nodes of contact, points of convergence, grey zones of negotiation, shared references and overlapping understandings of the world. This is one of the main ways in which cognitive worlds expand, and fully realising that cultural boundaries are far from absolute means leaving the Lévi-Straussian world of radical cultural difference and sharply bounded cultures behind once and for all.

In the misty dawn of time, when the world was still young and legible for the common people – say, three decades ago – not only did it seem possible to understand global politics through the double lens of the Cold War and imperialism; it was also fairly uncontroversial to divide the world into discrete cultural regions; the human world seemed to be a composite of relatively distinct cultures. Thus, in Denmark, people spoke Danish, and the Danes were liberal Protestants with characteristic Danish physiognomies who exhibited a characteristic body language and had a general cultural liking for red sausages and pilsner beer; in Bangladesh, the Bengali Muslims had their own customs and traditions; all the tribes in Kenya had their particular, unique cultures and languages, and so on. According to the prevailing worldview at that time, there was not much contact between these cultures, although there was some exchange and mutual influence going on. Cultural contact, which developed through imperialism and the growth of global capitalism, missionary work, development aid, migration and the diffusion of modern institutions such as the modern nation-state and the capitalist labour market, gave rise to a set of problems related to the encounter of separate cultures, each with its own special internal logic. Often, the results were misunderstandings and conflicts, and fairly often the politically and economically stronger culture came to dominate the weaker. This was often referred to as cultural imperialism – a term which is today rarely used, though it was common in intellectual discourse a surprisingly short time ago.

This understanding of culture and cultural differences, which has been fundamental to European thinking ever since Romanticism and crucial both in nationalist ideologies and anthropological thought, appears old-fashioned and dated now. This change is mainly a result of the fact that the world has changed. Although cultures have never been completely isolated and devoid of contact with other cultures, and despite the fact that cultural isolation has often been exaggerated both by scholars and others, the possibility of cultural isolation has shrunk at an overheated pace since the late twentieth century. Both economy and politics have become globalised or transnational. Because of advances in communication technology, money, goods, people, ideas and power travel across the world at increasing speed. There has been an enormous growth in air traffic during the past 60 years, and airline fares continue to drop. Satellite television, the internet and related technologies have accelerated the development of a world of communication without delays, where some events can be known about simultaneously everywhere, and where distances are shrinking rapidly.

Needless to say, changes of this magnitude have repercussions across people's life-worlds, and consequently new ways of communicating across cultural differences are required. It is still a common view that the processes of globalisation will lead to the annihilation of cultural differences and that human beings throughout the world are becoming more and more similar. This view is shared by both optimists who believe that poverty will be eradicated and democracy will eventually prevail worldwide, and by cultural pessimists who complain that the great cultural diversity of the planet is all but gone. A different perspective, common among intellectuals in the Global South, focuses on the neocolonial aspects of globalisation, emphasising how economic differences are sustained, and how the seemingly boundless openness that has characterised the era of globalisation is limited to the privileged.

A third approach involves the new forms of cultural variation that are evolving in the context of global modernity, because modernity does not equal cultural homogeneity. The globalisation of culture does not create global people. But globalisation creates 'cultural creoles', people who live at the intersection of different cultural traditions, constantly bombarded with impulses, expectations, demands and opportunities from several different angles, and who continuously create themselves, not from ready-made prescriptions but by crafting their own unique, complex cultural fabric (Hannerz 1990; see also Stewart 2007; Cohen and Sheringham 2016).

The old map creates a world of cultural islands (Eriksen 1993). On each island, people have their unique way of living, with their own traditions and so forth, but there is relatively little contact between the islands. It is extremely difficult to navigate the world of today using such a map. That is a main reason that a significant part of the academic community has been at work revising that map for some time. (Naturally, it is also possible to change the territory so that it matches the old map – a solution which might take the form of ethnic cleansing.) This revision process is evident in the bifurcation of history into different narratives – stories or histories. New nuances and a new diversity are incorporated into the conception of national identities, and new groups of people attain a sense of subjective and objective belonging to their society. Rebuilding the ship at sea has become a necessary task.

A kind of competence necessary for this task to be successful is, arguably, the ability to listen, a resource in notoriously short supply in the contemporary world. A cosmopolitan ethics may be a starting point, one contributing simultaneously to decolonising the minds of previously colonised peoples, and to bridging the gaps of intercultural relations through forms of communication where the symbolic power has been decentralised. Enlightenment philosophers, from Voltaire to Kant, envisioned a shrunken world, or at least one in which people of different cultural backgrounds were capable of living together.

In a review of Kwame Anthony Appiah's *Cosmopolitanism* (Appiah 2006), John Gray states that: 'As a position in ethical theory, cosmopolitanism is distinct from relativism and universalism. It affirms the possibility of mutual understanding between adherents to different moralities but without holding out the promise of any ultimate consensus' (Gray 2006).

In other words, fervent missionary activity is not, according to this view, compatible with a cosmopolitan outlook, nor is an ethical position which assumes that there is but one good life. These two initial principles are, incidentally, in line with Kant's view of cosmopolitanism, which emphasises the need to communicate across cultural and political boundaries, to accept hospitality when it is offered, and to respect the difference of the other without succumbing to relativist confusion. Precisely because overheating processes now are planetary and epidemic, the conditions for a genuinely global conversation look better than at any time in the past.

In this kind of world, irreducibly diverse and chronically overheated, we are all strangers in a strange land. At the same time, we are all in the same boat, divided by a shared destiny.

Bibliography

Aamnes, Frida (2015) Livsstil eller business? Bananbønder i en global økonomi (Way of life or business? Banana farmers in a global economy). MA dissertation, Department of Social Anthropology, University of Oslo.

Aarset, Monica Five (2014) *Hearts and Roofs: Family, Belonging, and (Un-) Settledness among Descendants of Immigrants in Norway*. PhD dissertation, Department of Social Anthropology, University of Oslo.

Aas, Katja Franko and Mary Bosworth, eds (2013) *The Borders of Punishment: Migration, Citizenship and Social Exclusion*. Oxford: Oxford University Press.

Adams, Douglas (1979) *The Hitchhiker's Guide to the Galaxy*. London: Pan.

Aguiar, José Carlos G. (2012) 'They come from China': Pirate CDs in Mexico in transnational perspective. In Gordon Mathews, Gustavo Lins Ribeiro and Carlos Alba Vega, eds, *Globalization from Below: The World's Other Economy*, pp. 36–53. London: Routledge.

Albrecht, Glenn (2005) Solastalgia, a new concept in human health and identity. *Philosophy Activism Nature* 3: 41–44.

Albrecht, Glenn et al. (2007) Solastalgia: The distress caused by environmental change. *Australasian Psychiatry* 15 (s1): S95–S98.

Alexander, Catherine (2009) Illusions of freedom: Polanyi and the third sector. In Chris Hann and Keith Hart, eds, *Market and Society: The Great Transformation Today*, pp. 221–239. Cambridge: Cambridge University Press.

Alexander, Catherine and Joshua Reno, eds (2012) *Economies of Recycling: The Global Transformation of Materials, Values and Social Relations*. London: Zed.

Anderson, Benedict (1992) *Long-distance Nationalism: World Capitalism and the Rise of Identity Politics*. Amsterdam: Centre for Asian Studies.

Andersson, Ruben (2014) *Illegality, Inc.: Clandestine Migration and the Business of Bordering Europe*. Berkeley: University of California Press.

Appiah, Kwame Anthony (2006) *Cosmopolitanism: Ethics in a World of Strangers*. New York: Norton.

Baer, Hans and Merrill Singer (2014) *The Anthropology of Climate Change: An Integrated Critical Perspective*. London: Routledge.

Ballard, J.G. (1982) *Myths of the Near Future*. London: Jonathan Cape.

Bandyopadhy, Ritayoti (2012) In the shadow of the mall: Street hawking in global Calcutta. In Gordon Mathews, Gustavo Lins Ribeiro and Carlos Alba Vega, eds, *Globalization from Below: The World's Other Economy*, pp. 171–185. London: Routledge.

Bangstad, Sindre (2014) *Anders Breivik and the Rise of Islamophobia*. London: Zed.

Barai, Munim K. (2012) Development dynamics of remittances in Bangladesh. Sage Open. DOI: 10.1177/2158244012439073

Barber, Benjamin (1995) *Jihad versus McWorld: How Globalism and Tribalism are Reshaping the World*. New York: Ballantine.
Barnes, Jessica and Michael Dove, eds (2015) *Climate Cultures: Anthropological Perspectives on Climate Change*. New Haven, CT: Yale University Press.
Barth, Fredrik (1961) *Nomads of South Persia: The Basseri Tribe of the Khamseh Confederacy*. Oslo: Universitetsforlaget.
——(1975) *Ritual and Knowledge among the Baktaman*. Oslo: Universitetsforlaget.
Bateson, Gregory (1972) *Steps to an Ecology of Mind*. New York: Chandler.
——(1978) *Mind and Nature*. Glasgow: Fontana.
Bateson, Gregory, Donald Jackson, Jay Haley and John Weakland (1956) Toward a theory of schizophrenia. *Behavioral Science* 1: 251–264.
Bauman, Zygmunt (1999) *Globalization: The Human Consequences*. New York: Columbia University Press.
——(2004) *Wasted Lives: Modernity and its Outcasts*. Cambridge: Polity.
Bear, Laura (2014) For labour: Ajeet's accident and the ethics of technological fixes in time. *Journal of the Royal Anthropological Institute* 20: 71–88.
Beck, Ulrich (1992) *Risk Society: Towards a New Modernity*. Cambridge: Polity.
Becker, Elizabeth (2013) *Overbooked: The Exploding Business of Travel and Tourism*. New York: Simon & Schuster.
Behrends, Andrea, Stephen Reyna and Günther Schlee, eds (2011) *Crude Domination: An Anthropology of Oil*. Oxford: Berghahn.
Bendixsen, Synnøve (2015) 'Give me the damn papers!' Å vente på oppholdstillatelse ('Give me the damn papers!' Waiting for a permit of residence). *Norsk antropologisk tidsskrift* 26: 285–303.
Berreman, Gerald (1978) Scale and social relations. *Current Anthropology*, 19(2): 225–245.
Blommaert, Jan (2010) *The Sociolinguistics of Globalization*. Cambridge: Cambridge University Press.
Bloomberg (2015) How many text messages are sent each year? http://www.bloomberg.com/news/videos/b/443bcbc2-2d56-4608-a2e8-732689593f17, accessed 13 October 2015.
Boyer, Dominic (2014) Energopower: An introduction, *Anthropological Quarterly* 87(2): 309–334.
Caldeira, Teresa P.R. (2001) *City of Walls: Crime, Segregation, and Citizenship in São Paulo*. Berkeley: University of California Press.
Calhoun, Craig (2002) The class consciousness of frequent travellers: Towards a critique of actually existing cosmopolitanism. In S. Vertovec and R. Cohen, eds, *Conceiving Cosmopolitanism: Theory, Context and Practice*, pp. 86–109. Oxford: Oxford University Press.
Carr, Nicholas (2008) Is Google making us stupid? What the Internet is doing to our brains. *The Atlantic*, July/August.
Carroll, Lewis (2003 [1872]) *Through the Looking Glass*. London: Penguin Popular Classics.
Carson, Rachel (1962) *Silent Spring*. Boston, MA: Houghton Mifflin.
Castells, Manuel (1996) *The Rise of the Network Society*, vol. 1: *The Information Age: Economy, Society and Culture*. Oxford: Blackwell.

—— (1997) *The Power of Identity*, vol. 2: *The Information Age: Economy, Society and Culture*. Oxford: Blackwell.

—— (1998) *End of the Millennium*, vol. 3: *The Information Age: Economy, Society and Culture*. Oxford: Blackwell.

Castles, Stephen and Alasdair Davidson (2000) *Citizenship and Migration: Globalization and the Politics of Belonging*. London: Palgrave Macmillan.

Center for Biological Diversity (2015) Species extinction and human population. http://www.biologicaldiversity.org/programs/population_and_sustainability/extinction_and_population_graph.html, accessed 23 September 2015.

Clastres, P. (1987 [1974]) *Society Against the State: Essays in Political Anthropology*. New York: Zone Books.

Clemens, Michael A. and David McKenzie (2014) Why don't remittances appear to affect growth? Working Paper 366. Washington, DC: Center for Global Development.

Cohen, Robin and Olivia Sheringham (2016) *Encountering Difference*. Cambridge: Polity.

Comaroff, John L. and Jean Comaroff (2009) *Ethnicity, Inc.* Chicago: University of Chicago Press.

Crate, Susan A. and Mark Nuttall, eds (2009) *Anthropology and Climate Change: From Encounters to Actions*. Walnut Creek, CA: Left Coast Press.

Davis, Mike (2006) *Planet of Slums*. London: Verso.

Deacon, Terrence W. (2012) *Incomplete Nature: How Mind Emerged from Matter*. New York: W.W. Norton.

Douglas, Mary (1966) *Purity and Danger: An Analysis of Concepts of Pollution and Taboo*. London: Routledge & Kegan Paul.

—— (1992) *Risk and Blame: Essays in Cultural Theory*. London: Routledge.

Edvardsen, Cathrine Heisholt (2006) Internett i afrikanske skoler? (Internet in African schools?) In Thomas Hylland Eriksen, ed., *Internett i praksis* (Internet in practice), pp. 93–110. Oslo: Spartacus.

Eide, Elisabeth, Risto Kunelius and Angela Phillips, eds (2008) *Transnational Media Events: The Mohammed Cartoons and the Imagined Clash of Civilizations*. Göteborg: Nordicom.

Erem, Suzan and E. Paul Durrenberger (2008) *On the Global Waterfront: The Fight to Free the Charleston 5*. New York: Monthly Review Press.

Eriksen, Thomas Hylland (1993) Do cultural islands exist? *Social Anthropology* 2(1): 133–147.

—— (2001) *Tyranny of the Moment: Fast and Slow Time in the Information Age*. London: Pluto.

—— (2004) *Røtter og* føtter: Identitet i en omskiftelig tid (Roots and feet: Identity at an unstable time'). Oslo: Aschehoug.

—— (2005) Mind the gap: Flexibility, epistemology and the rhetoric of new work. *Cybernetics and Human Knowing* 12(1–2): 50–60.

—— (2007) Stacking and continuity: On temporal regimes in popular culture. In Robert Hassan and Ronald E. Purser, eds, *24/7: Time and Temporality in the Network Society*, pp. 141–160. Stanford, CA: Stanford University Press.

—— (2008) *Storeulvsyndromet: Jakten på lykken i overflodssamfunnet.* (The syndrome of the big, bad wolf: The search for happiness in the affluent society). Oslo: Aschehoug.
——(2010) *Samfunn* (Society). Oslo: Universitetsforlaget.
——(2011) *Søppel: Avfall i en verden av bivirkninger* (Rubbish: Waste in a world of side-effects). Oslo: Aschehoug.
——(2014a) *Globalization: The Key Concepts*, 2nd edn. London: Bloomsbury.
——(2014b) Who or what to blame: Competing interpretations of the Norwegian terrorist attack. *Archives European Journal of Sociology* 55: 275–294.
——(2015) *Fredrik Barth: An Intellectual Biography*. London: Pluto.
——(forthcoming a) Conflicting regimes of knowledge about Gladstone harbour: A drama in four acts.
——(forthcoming b) Knowledge and power around the East End Mine.
Evans-Pritchard, E.E. (1983 [1937]) *Witchcraft, Magic and Oracles among the Azande*, ed. Eva Gillies. Oxford: Oxford University Press.
Fabian, Johannes (1983) *Time and the Other: How Anthropology Makes Its Object.* New York: Columbia University Press.
Ferguson, James (1999) *Expectations of Modernity: Myth and Meanings of Urban Life on the Zambian Copperbelt.* Berkeley: University of California Press.
Fukuyama, Francis (1995) *Trust: The Social Virtues and the Creation of Prosperity.* London: Penguin.
Furedi, Frank (2002) *Culture of Fear: Risk-taking and the Morality of Low Expectations.* London: Continuum.
Gaarder, Jostein (2007) Et kosmisk ansvar (A cosmic responsibility). *Aftenposten*, 9 June.
Gauthier, Mélissa (2012) Mexican 'ant traders' in the El Paso/Ciudad Juarez border region: Tensions between globalization, securitization and new mobility regimes. In Gordon Mathews, Gustavo Lins Ribeiro and Carlos Alba Vega, eds, *Globalization from Below: The World's Other Economy*, pp. 138–153. London: Routledge.
Geertz, Clifford (1957) Ritual and social change: A Javanese example. *American Anthropologist* 59(1): 32–54.
——(1986) The uses of diversity. *Michigan Quarterly Review* 25(1): 105–123.
Gellner, Ernest (1983) *Nations and Nationalism*. Oxford: Blackwell.
——(1988) Trust, cohesion, and the social order. In Diego Gambetta, ed., *Trust: Making and Breaking Cooperative Relations*, pp. 142–157. Oxford: Blackwell.
Giddens, Anthony (1990) *The Consequences of Modernity*. Cambridge: Polity.
Glackin, Shane N. (2011) Kind-making, objectivity, and political neutrality: The case of Solastalgia. *Studies in History and Philosophy of Biological and Biological and Biomedical Sciences* 43: 209–218.
Glick Schiller, Nina, Linda Basch and Cristina Szanton Blanc (1995) From immigrant to transmigrant: Theorizing transnational migration. *Anthropological Quarterly* 68 (1): 48–63.
Golub, Alex (2014) *Leviathans at the Gold Mine: Creating Indigenous and Corporate Actors in Papua New Guinea.* Durham, NC: Duke University Press.

Goody, Jack (1977) *The Domestication of the Savage Mind*. Cambridge: Cambridge University Press.

Gould, Stephen Jay (2002) *The Structure of Evolutionary Theory*. Cambridge, MA: Belknap.

Gould, Stephen Jay and Elizabeth Vrba (1981) Exaptation: A missing term in the science of form. *Paleobiology* 8: 4–15.

Grantham, Jeremy (2011) Time to wake up: Days of abundant resources and falling prices are over forever. *The Oil Drum*, http://www.theoildrum.com/node/7853, accessed 12 August 2015.

Gray, John (1998) *False Dawn: The Delusions of Global Capitalism*. London: Granta.

—— (2006) Easier said than done. *The Nation*, 30 January.

Grønhaug, Reidar (1978) Scale as a variable in analysis: Fields in social organization in Herat, northwest Afghanistan. In Fredrik Barth, ed., *Scale and Social Organization*, pp. 78–121. Oslo: Universitetsforlaget.

Günel, Gökce (2011) Spaceship in the desert: Abu Dhabi's Masdar City. *Anthropology News* October: 3–4.

Gupta, Akhil (2015) An anthropology of electricity from the Global South. *Global Anthropology* 30(4): 555–568.

Hall, Alexandra and Lisette Josephides, eds (2013) *We the Cosmopolitans*. Oxford: Berghahn.

Hann, Chris, ed. (1994) *When History Accelerates: Essays on Rapid Social Change, Complexity and Creativity*. London: Athlone.

Hann, Chris and Keith Hart (2011) *Economic Anthropology*. Cambridge: Polity.

Hannerz, Ulf (1990) Cosmopolitans and locals in world culture. In Mike Featherstone, ed., *Global Culture: Nationalism, Globalization and Modernity*, pp. 237–52. London: Sage.

Harris, Marvin (1978) *Cannibals and Kings: The Origins of Culture*. Glasgow: Fontana.

Harris, Nigel (2002) *Thinking the Unthinkable: The Immigrant Myth Exposed*. London: I.B. Tauris.

Harrison, Simon (2000) Identity as a scarce resource. *Social Anthropology* 7(3): 239–252.

Hart, Keith (1973) Informal income opportunities and urban employment in Ghana. *Journal of Modern African Studies* 11(1): 61–89.

—— ed. (2015) *Economy For and Against Democracy*. Oxford: Berghahn.

Hart, Keith, Jean-Louis Laville and Antonio David Cattani, eds (2010) *The Human Economy*. Cambridge: Polity.

Harvey, David (2005) *A Short History of Neoliberalism*. Oxford: Oxford University Press.

Hassan, Robert and Ronald Purser, eds (2007) *24/7: Time and Temporality in the Network Society*. Stanford, CA: Stanford Business Books.

Hastrup, Kirsten and Karen Fog Olwig, eds (2012) *Climate Change and Human Mobility: Challenges to the Social Sciences*. Cambridge: Cambridge University Press.

Hendry, Joy (2014) *Science and Sustainability: Learning from Indigenous Wisdom*. Basingstoke: Palgrave Macmillan.

Hessen, Dag O. and Thomas Hylland Eriksen (2012) *På stedet løp: Konkurransens paradokser* (Moving still: The paradoxes of competition). Oslo: Aschehoug.

Hobsbawm, Eric (1994) *Age of Extremes: The Short Twentieth Century 1914–1991*. London: Abacus.

Homer-Dixon, Thomas (2006) *The Upside of Down: Catastrophe, Creativity, and the Renewal of Civilization*. Washington, DC: Island Press.

Hornborg, Alf (2011) *Global Ecology and Unequal Exchange: Fetishism in a Zero-Sum World*. London: Routledge.

Horst, Heather (2006) The blessings and burdens of communication: Cell phones in Jamaican transnational fields. *Global Networks* 6(2): 143–159.

Horst, Heather and Daniel Miller (2006) *The Cell Phone: An Anthropology of Communication*. Oxford: Berg.

—— eds (2012) *Digital Anthropology*. Oxford: Berg.

Howell, Signe (2015) Politics of appearances: Some reasons why the UN-REDD project in Central Sulawesi failed to unite the various stakeholders. *Asia Pacific Viewpoint* 56(1): 37–47.

Huntington, Samuel (1996) *The Clash of Civilizations and the Remaking of a World Order*. New York: Simon & Schuster.

ILO (2014) *Transitioning from the Informal to the Formal Economy*. Report V(1). Geneva: International Labour Organization.

ITU (International Telecommunications Union) (2015) ICT Facts and Figures 2015. http://www.itu.int/en/ITU-D/Statistics/Pages/stat/default.aspx, accessed 14 February 2016.

Jackson, Deborah Davis (2011) Scents of place: The dysplacement of a First Nations community in Canada. *American Anthropologist* 113(4): 606–618.

Kasten, Erich, ed. (2004) *Properties of Culture - Culture as Property*. Berlin: Dietrich Reimer Verlag.

Kirsch, Stuart (2014) *Mining Capitalism: The Relationship between Corporations and their Critics*. Berkeley: University of California Press.

Knudsen, Are (2015) De er flyktninger 2.0 (They are refugees 2.0). *Bergens Tidende*, 22 June.

Lazar, Sian, ed. (2014) *The Anthropology of Citizenship: A Reader*. London: Wiley.

Lévi-Strauss, Claude (1961 [1955]) *Tristes tropiques*, trans. John Russell. New York: Criterion.

—— (1966 [1962]) *The Savage Mind*. Chicago: University of Chicago Press.

—— (1969 [1949]) *The Elementary Structures of Kinship*, trans. Rodney Needham. London: Tavistock.

Levinson, Marc (2006) *The Box: How the Shipping Container Made the World Smaller and the World Economy Bigger*. Princeton, NJ: Princeton University Press.

Lorenz, Edward (1972) Deterministic nonperiodic flow. *Journal of Atmospheric Sciences* 20: 130–141.

Lovelock, James (2006) *The Revenge of Gaia: Why the Earth Is Fighting Back – and How We Can Still Save Humanity*. London: Allen Lane.

Lucke, Alec (2013) *Road to Exploitation: Political Capture by Mining in Queensland.* Gladstone: Gladstone Printing Services.

Luning, Sabine and Robert J. Pijpers (forthcoming) Governing access to gold in Ghana: In-depth geopolitics on mining concessions. *Africa.*

Maeckelbergh, Marianne (2009) *The Will of the Many: How the Alterglobalisation Movement is Changing the Face of Democracy.* London: Pluto.

Malinowski, Bronislaw (1984 [1922]) *Argonauts of the Western Pacific.* Prospect Heights, IL: Waveland.

Mandelbrot, Benoît (1967) How long is the coast of Britain? Statistical self-similarity and fractional dimension. *Science* n.s., 156: 636–638.

Mason, Paul (2015) *Postcapitalism: A Guide to Our Future.* London: Allen Lane.

Mathews, Gordon (2012) Neoliberalism and globalization from below in Chungking Mansions, Hong Kong. In Gordon Mathews, Gustavo Lins Ribeiro and Carlos Alba Vega, eds, *Globalization from Below: The World's Other Economy*, pp. 69-85. London: Routledge.

Mathews, Gordon and Carlos Alba Vega (2012) Introduction: What is globalization from below? In Gordon Mathews, Gustavo Lins Ribeiro and Carlos Alba Vega, eds, *Globalization from Below: The World's Other Economy*, pp. 1–16. London: Routledge.

Mathews, Gordon, Gustavo Lins Ribeiro and Carlos Alba Vega, eds (2012) *Globalization from Below: The World's Other Economy.* London: Routledge.

McCarthy, Tom (2015) *Satin Island.* London: Jonathan Cape.

Mead, Margaret (2002 [1956]) *New Lives for Old: Cultural Transformation in Manus, 1928–1953.* New York: HarperCollins.

Meadows, Donella H., Dennis L. Meadows, Jørgen Randers and William W. Behrens III (1972) *The Limits to Growth.* New York: Universe Books.

Mintz, Sidney W. (1953) The culture history of a Puerto Rican sugar cane plantation: 1876–1949. *American Historical Review* 33(2): 224–251.

Mitchell, Timothy (2011) *Carbon Democracy: Political Power in the Age of Oil.* London: Verso.

Moen, Ron A., John Pastor and Yosef Cohen (1999) Antler growth and extinction of Irish elk. *Evolutionary Ecology Research* 1: 235–249.

Monbiot, George (2014) The impossibility of growth. *The Guardian*, 27 May.

Morozov, Evgeny (2012) *The Net Delusion: How Not to Liberate the World.* London: Penguin.

Nadel, Siegfried (1952) Witchcraft in four African societies: An essay in comparison. *American Anthropologist* 54(1): 18–29.

NEF (New Economics Foundation) (2014) The Happy Planet Index 2014. http://www.happyplanetindex.org/data/, accessed 10 August 2015.

New York Times (2015) Photos, photos everywhere. 23 July.

Norgaard, Kari Marie (2011) *Living in Denial: Climate Change, Emotions, and Everyday Life.* Cambridge, MA: MIT Press.

Nørretranders, Tor (1999) *The User Illusion.* London: Penguin.

OECD (Organisation for Economic Co-operation and Development) (2015) Carbon dioxide emissions embedded in international trade. http://www.

oecd.org/sti/ind/carbondioxideemissionsembodiedininternationaltrade.htm, accessed 3 August 2015.
Ogburn, William F. (1922) *Social Change with Respect to Culture and Original Nature*. New York: B.W. Huebsch.
Ose, Tommy (2016) *Food Waste in Northern Norway*. PhD dissertation, Department of Social Anthropology, University of Oslo.
Ostrom, Elinor (1990) *Governing the Commons: The Evolution of Institutions for Collective Action*. Cambridge: Cambridge University Press.
Our Finite World (2015) World energy consumption since 1820 in charts. https://ourfiniteworld.com/2012/03/12/world-energy-consumption-since-1820-in-charts/, accessed 25 August 2015.
Pariser, Eli (2011) *The Filter Bubble: What the Internet is Hiding From You*. New York: Penguin.
Pijpers, Robert J. (2014). Crops and carats: Exploring the interconnectedness of mining and agriculture in Sub-Saharan Africa. *Futures* 62(A): 32–39.
—— (forthcoming). Lost glory or poor legacy? Mining pasts, future projects in rural Sierra Leone. In J.B. Gewald, J. Jansen and S. Luning, eds, *Mining History: Corporate Strategies, Heritage and Development*. London: Routledge.
Pliez, Olivier (2012) Following the new Silk Road between Yiwu and Cairo. In Gordon Mathews, Gustavo Lins Ribeiro and Carlos Alba Vega, eds, *Globalization from Below: The World's Other Economy*, pp. 19–35. London: Routledge.
Polanyi, Karl (1957 [1944]) *The Great Transformation: The Political and Economic Origins of our Time*. Boston, MA: Beacon Press.
Puiu, Tibi (2015) Not even World War III will stop unsustainable human population growth. *ZME Science*. http://www.zmescience.com/science/unsustainable-human-population-growth-0534, accessed 14 February 2016.
Rakopoulos, Theoodoros (2014) Resonance of solidarity: Meaning of a local concept in anti-austerity Greece. *Journal of Modern Greek Studies* 32(2): 313–337.
Rappaport, Roy A. (1968) *Pigs for the Ancestors: Ritual in the Ecology of a New Guinea People*. New Haven, CT: Yale University Press.
Rathje, William and Cullen Murphy (2001) *Rubbish! The Archaeology of Garbage*. Tucson: University of Arizona Press.
Reno, Joshua (2009). Your trash is someone's treasure: The politics of value at a Michigan landfill. *Journal of Material Culture* 14(1): 29–46.
Reyna, Stephen and Andrea Behrends (2008) The crazy curse and the crude domination: Towards an anthropology of oil. *Focaal* 52: 3–15.
Ribeiro, Gustavo Lins (2006) Economic globalization from below. *Etnográfica* 10(2): 233–249.
Richerson, Peter (2014) Comment on 'Blueprint for the global village'. *Cliodynamics* 5: 136–140.
Ridley, Matt (1993) *The Red Queen: Sex and the Evolution of Human Nature*. London: Penguin.
—— (1996) *The Origins of Virtue*. London: Viking.
Rifkin, Jeremy (2011) *The Third Industrial Revolution: How Lateral Power is Transforming Energy, the Economy, and the World*. London: Palgrave Macmillan.

de Rijke, Kim (2013) Hydraulically fractured: Unconventional gas and anthropology. *Anthropology Today* 29(2): 13–17.

Ritzer, George (2004) *The Globalization of Nothing*. London: Sage.

Røhnebæk, Maria (2006) Hvis internett er svaret, hva er spørsmålet? (If internet is the answer, what is the question?) In Thomas Hylland Eriksen, ed., *Internett i praksis* (Internet in practice), pp. 111–140. Oslo: Spartacus.

Rosa, Hartmut (2013) *Social Acceleration: A New Theory of Modernity*. New York: Columbia University Press.

Rose, Steven (1996) *Lifelines: Biology Beyond Determinism*. Oxford: Oxford University Press.

Rothstein, Bo (2000) Trust, social dilemmas, and collective memories: On the rise and decline of the Swedish model. *Journal of Theoretical Politics* 12: 477–499.

Rowan, Michael (2014) We need to talk about growth. *Persuade Me*, http://persuademe.com.au/need-talk-growth-need-sums-well/, accessed February 2016.

Sahlins, Marshall D. (1968) Notes on the original affluent society. In Richard B. Lee and Irven DeVore, eds, *Man the Hunter*, pp. 85–89. Chicago: Aldine.

—— (1994) Goodbye to tristes tropes: Ethnography in the context of modern world history. In Robert Borofsky, ed., *Assessing Cultural Anthropology*, pp. 377–394. New York: McGraw-Hill.

Sassen, Saskia (2014) *Expulsions: Brutality and Complexity in the Global Economy*. Cambridge, MA: Belknap Press.

Schober, Elisabeth (2015) Making a living, making life worth living: Filipino scavengers and Korean shipbuilders in Subic Bay. Paper presented at Overheating workshop 'Waste and the superfluous', University of Oslo, 14–15 September.

Schumacher, E.F. (1973) *Small Is Beautiful: A Study of Economics as if People Mattered*. London: Blond & Briggs.

Scott, James C. (1998) *Seeing Like a State: How Certain Schemes to Improve the Human Condition have Failed*. New Haven, CT: Yale University Press.

Self, Will (2014) The novel is dead (this time it's for real). *The Guardian*, 2 May.

Sennett, Richard (1998) *The Corrosion of Character: Personal Consequences of Work in the New Capitalism*. New York: W.W. Norton.

Shore, Cris and Susana Trnka, eds (2013) *Up Close and Personal: On Peripheral Perspectives and the Production of Anthropological Knowledge*. Oxford: Berghahn.

Shrady, Nicholas (2009) *The Last Day: Wrath, Ruin, and Reason in the Great Lisbon Earthquake of 1755*. London: Penguin.

Simmel, Georg (1990 [1900]) *The Philosophy of Money*. London: Routledge.

Smith, John Maynard and Eörs Szathmáry (1995) *The Major Transitions in Evolution*. Oxford: Oxford University Press.

Smithsonian Institution (2015) When will we hit peak garbage? http://www.smithsonianmag.com/science-nature/when-will-we-hit-peak-garbage-7074398/?no-ist), accessed 15 August 2015.

Sørhaug, Christian (2012) Holding house in crazy waters: An exploration of householding practices among the Warao, Orinoco Delta, Venezuela. PhD dissertation, Department of Social Anthropology, University of Oslo.
Soros, George (2002) *George Soros on Globalization*. Oxford: Public Affairs.
de Soto, Hernando (2007) *The Mystery of Capital: Why Capitalism Triumphs in the West and Fails Everywhere Else*. New York: Basic Books.
Sperber, Dan and Deirdre Wilson (1986) *Relevance: Communication and Cognition*. Oxford: Blackwell.
Stade, Ronald (2013) The politics of human waste in Accra, Ghana. Seminar paper, 'Overheating', University of Oslo, 27 August.
Standing, Guy (2011) *The Precariat: The New Dangerous Class*. London: Bloomsbury Academic.
Statistics Norway (2015) Innvandrere og norskfødte med innvandrerforeldre 1970–2014 (Immigrants and Norwegian-born with immigrant parents 1970–2014). https://www.ssb.no/innvandring-og-innvandrere/nokkeltall, accessed 15 August 2015.
Steffen, Will, Paul J. Crutzen and John R. McNeill (2007) The Anthropocene: Are human beings now overwhelming the forces of nature? *AMBIO* 36(8): 614–621.
Steinert, Margrethe (2015) Change as continuity; continuity through change: An anthropological study of the Asháninka of the Peruvian highland Amazonia. MA dissertation, Department of Social Anthropology, Oslo University.
Stewart, Charles, ed. (2007) *Creolization: History, Ethnography, Theory*. Walnut Creek, CA: Left Coast Press.
Stiglitz, Joseph (2002) *Globalization and its Discontents*. London: Allen Lane.
Tainter, Joseph A. (1988) *The Collapse of Complex Societies*. Cambridge: Cambridge University Press.
——(2014) Collapse and sustainability: Rome, the Maya, and the modern world. *Archaeological Papers of the American Anthropological Association* 24: 201–214.
Thompson, Michael (1979) *Rubbish Theory: The Creation and Destruction of Value*. Oxford: Oxford University Press.
Thorleifsson, Cathrine Moe (forthcoming) From Coal to UKIP: Fuelling nationalism in post-industrial Doncaster.
Trawick, Paul and Alf Hornborg (2015) Revisiting the image of limited good: On sustainability, thermodynamics, and the illusion of creating wealth. *Current Anthropology* 56(1): 1–27.
Trouillot, Michel-Rolph (2001) Close encounters of the deceptive kind. *Current Anthropology* 42: 125–138.
Tsing, Anna Lowenhaupt (2012) On nonscalability: The living world is not amenable to precision-nested scales. *Common Knowledge* 18(3): 505–524.
Turner, Graham (2014) Is global collapse imminent? MSSI Research Paper 4, Melbourne Sustainable Society Institute, The University of Melbourne.
Uimonen, Paula (2001) Transnational.dynamics@development.net: Internet, modernization and globalization. PhD dissertation, Department of Social Anthropology, Stockholm University.

UN (1987) *Our Common Future: Report of the World Commission on Environment and Development*. New York: United Nations.

—— (2014) World Urbanization Prospects, 2014 Update. http://esa.un.org/unpd/wup/highlights/wup2014-highlights.pdf, accessed 29 July 2015.

UNHCR (United Nations High Commissioner for Refugees) (2013) Global forced displacement tops 50 million for first time in post-World War II era. http://www.unhcr.org/53a155bc6.html, accessed 5 August 2015.

—— (2016) Syria regional refugee response. http://data.unhcr.org/syrianrefugees/regional.php, accessed 5 August 2015.

UNWTO (United Nations World Tourism Organization) (2014) *Tourism Highlights*, 2014 edition. Madrid: UNWTO.

Urry, John (2003) *Global Complexity*. Cambridge: Polity.

Van Valen, Leigh (1973) A new evolutionary law. *Evolutionary Theory* 1: 1–30.

Vertovec, Steven (2004) Cheap calls: the social glue of migrant transnationalism. *Global Networks* 4(2): 219–224.

—— (2007) Super-diversity and its implications. *Ethnic and Racial Studies* 30(6): 1024–1054.

Vindegg, Mikkel (2015) The power to produce: The influence of limited electricity access in a Nepali textile industry. MA dissertation, Department of Social Anthropology, University of Oslo.

Virilio, Paul (2000) *The Information Bomb*. London: Verso.

Wajcman, Judy (2015) *Pressed for Time: The Acceleration of Life in Digital Capitalism*. Chicago: University of Chicago Press.

Wengrow, David and David Graeber (2015) Farewell to the 'childhood of man': Ritual, seasonality, and the origins of inequality. *Journal of the Royal Anthropological Institute* 21(3): 597–619.

Wessendorf, Susanne (2014) *Commonplace Diversity: Social Relations in a Super-Diverse Context*. London: Palgrave.

White, Leslie (1943) Energy and the evolution of culture. *American Anthropologist* 45(3): 335–356.

—— (1949) *The Science of Culture: A Study of Man and Civilization*. New York: Farrar, Straus & Giroux.

—— (1959) *The Evolution of Culture: The Development of Civilization to the Fall of Rome*. New York: McGraw-Hill.

Wilhite, Harold (2012) A socio-cultural analysis of changing household electricity consumption in India. In Daniel Spreng et al., eds, *Tackling Long-term Global Energy Problems*. Environment and Policy vol. 52. Dordrecht: Springer, pp. 97–113.

Wilson, David Sloan and Dag O. Hessen (2014) Blueprint for the global village. *Cliodynamics* 5: 123–157.

Winther, Tanja (2008) *The Impact of Electricity: Development, Desires and Dilemmas*. Oxford: Berghahn.

Wordpress (2015) Some interesting facts about blogs. http://www.wpvirtuoso.com/how-many-blogs-are-on-the-internet/, accessed 13 October 2015.

World Bank (2014) Carbon emissions per capita. http://data.worldbank.org/indicator/EN.ATM.CO2E.PC, accessed 10 August 2015.

Index

Charts and diagrams are indicated by 'fig' following the page number.